The Coroners of Northern Britain c. 1300–1700

DOI: 10.1057/9781137381071.0001

Other Palgrave Pivot titles

Christina Slade: Watching Arabic Television in Europe: From Diaspora to Hybrid Citizens

Fred E. Knowles: The Indian Law Legacy of Thurgood Marshall

Louisa Hadley: Responding to Margaret Thatcher's Death

Kylie Mirmohamadi: The Digital Afterlives of Jane Austen: Janeites at the Keyboard

Rebeka L. Maples: The Legacy of Desegregation: The Struggle for Equality in Higher Education

Stijn Vanheule: Diagnosis and the DSM: A Critical Review

James DeShaw Rae: Analyzing the Drone Debates: Targeted Killing, Remote Warfare, and Military Technology

Torben Bech Dyrberg: Foucault on the Politics of Parrhesia

Bernice M. Murphy: The Highway Horror Film

Jolene M. Sanders: Women in Narcotics Anonymous: Overcoming Stigma and Shame

Bruce E. Bechtol, Jr.: North Korea and Regional Security in the Kim Jong-un Era: A New International Security Dilemma

Patrick Alan Danaher, Andy Davies, Linda De George-Walker, Janice K. Jones, Karl J. Matthews, Warren Midgley, Catherine H. Arden, Margaret Baguley: Contemporary Capacity-Building in Educational Contexts

Margaret Baguley, Patrick Alan Danaher, Andy Davies, Linda De George-Walker, Janice K. Jones, Karl J. Matthews, Warren Midgley and Catherine H. Arden: Educational Learning and Development: Building and Enhancing Capacity

Marian Lief Palley and Howard A. Palley: The Politics of Women's Health Care in the United States

Nikhilesh Dholakia and Romeo V. Turcan: Toward a Metatheory of Economic Bubbles: Socio-Political and Cultural Perspectives

Tommi A. Vuorenmaa: Lit and Dark Liquidity with Lost Time Data: Interlinked Trading Venues around the Global Financial Crisis

Ian I. Mitroff, Can M. Alpaslan and Ellen S. O'Connor: Everybody's Business: Reclaiming True Management Skills in Business Higher Education

Helen Jefferson Lenskyj: Sexual Diversity and the Sochi 2014 Olympics: No More Rainbows

Laurence Pope: The Demilitarization of American Diplomacy: Two Cheers for Striped Pants

P. Carl Mullan: The Digital Currency Challenge: Shaping Online Payment Systems through US Financial Regulations

Ana María Relaño Pastor: Shame and Pride in Narrative: Mexican Women's Language Experiences at the U.S.–Mexico Border

Manohar Pawar: Water and Social Policy

Jennifer Yamin-Ali: Data-Driven Decision-Making in Schools: Lessons from Trinidad

Lionel Gossman: André Maurois (1885–1967): Fortunes and Misfortunes of a Moderate

Matthew Watson: Uneconomic Economics and the Crisis of the Model World

DOI: 10.1057/9781137381071.0001

palgrave▸**pivot**

The Coroners of Northern Britain c. 1300–1700

R. A. Houston

Professor of Early Modern History, University of St Andrews, UK

DOI: 10.1057/9781137381071.0001

First published 2014 by
PALGRAVE MACMILLAN

Palgrave Macmillan in the UK is an imprint of Macmillan Publishers Limited, registered in England, company number 785998, of Houndmills, Basingstoke, Hampshire RG21 6XS.

Palgrave Macmillan in the US is a division of St Martin's Press LLC, 175 Fifth Avenue, New York, NY 10010.

Palgrave Macmillan is the global academic imprint of the above companies and has companies and representatives throughout the world.

Palgrave® and Macmillan® are registered trademarks in the United States, the United Kingdom, Europe and other countries

ISBN: 978-1-137-38108-8 EPUB
ISBN: 978-1-137-38107-1 PDF
ISBN: 978-1-137-38106-4 Hardback

This book is printed on paper suitable for recycling and made from fully managed and sustained forest sources. Logging, pulping and manufacturing processes are expected to conform to the environmental regulations of the country of origin.

A catalogue record for this book is available from the British Library.

A catalog record for this book is available from the Library of Congress.

www.palgrave.com/pivot

DOI: 10.1057/9781137381071

But the crowner he cum and the justice too,
With a hue and a cry and a hullabaloo.

'The twa sisters', in F. J. Child (ed.), *The English and Scottish popular ballads*, 5 vols (London: H. Stevens, Son & Stiles, 1882–98), vol. 1, 141.

DOI: 10.1057/9781137381071.0001

Contents

DOI: 10.1057/9781137381071.0001

Acknowledgements

The research on which this book is based began during a Leverhulme Major Research Fellowship. I am most grateful to the Trustees. It grew out of curiosity about how Scots investigated suicide and what coroners actually did there, which arose while researching another book that I wrote during the fellowship: *Punishing the dead? Suicide, lordship, and community in Britain, 1500–1830* (Oxford: Oxford UP, 2010). I should like to thank the following for their help in preparing the present book: Jackson Armstrong, Geoffrey Barrow, Steve Boardman, Michael Brown, Andrew Burt, Barbara Crawford, Jon Crawford, John Finlay, Julian Goodare, Alexander Grant, John Harrison, John Hudson, Krista Kesselring, Peter McNeill, Jean Munro, Steve Murdoch, Athol Murray, Peter Murray, Cynthia Neville, Robert Shiels, Katie Stevenson, Keith Stringer, Alice Taylor, Michael Wasser, Cassie Watson and Alex Woolf.

DOI: 10.1057/9781137381071.0002

List of Abbreviations

APS   Acts of the Parliaments of Scotland
ATS   Accounts of the Treasurer of Scotland
BL   British Library
ERS   Exchequer Rolls of Scotland
NA   National Archives
NAS   National Archives of Scotland
NLS   National Library of Scotland
NRAS   National Register of Archives for Scotland
RMS   Register of the Great Seal of Scotland
RPCS   Register of the Privy Council of Scotland (first series unless otherwise stated)
UP   University Press

DOI: 10.1057/9781137381071.0003

A Note on Presentation

I have chosen to retain historic spellings and nomenclature to clarify differences. For example, itinerant royal courts in medieval Scotland were called 'ayres', those in England 'eyres'; the English termed commissioners of the peace 'Justices of the Peace' whereas their Scottish equivalents were 'Justices of Peace'. Possibly unfamiliar words in Scots, English, Gaelic, Welsh, legal French and Latin are explained in the text, but S. R. O'Rourke, *Glossary of legal terms* (Edinburgh: Thomson/W. Green, 2004) may also be useful. I have chosen to locate all place names I could identify within pre-1975 counties or shires. For historical maps of the main administrative areas of Scotland, see P. G. B. McNeill and H. L. MacQueen (eds), *Atlas of Scottish history to 1707* (Edinburgh: Scottish Medievalists and Department of Geography, University of Edinburgh, 1996). Finally, those who wish a crash course in Scottish history may find helpful my *Scotland: a very short introduction* (Oxford: Oxford UP, 2008).

Introduction: The History of Coroners in Britain

Abstract: *For the last 800 years, coroners have been important in England's legal and political landscape, best known as investigators of sudden, suspicious, violent or unnatural death. Transplanted early on to Wales and Ireland, the office is today thoroughly familiar, found in a recognisable form over much of the English-speaking world, notably in the United States, Canada, Australia and New Zealand. In contrast, historians have largely ignored, misunderstood or dismissed the office of coroner in Scotland. The introduction sets out what coroners did in historic Britain and their place in the different judicial, political and social systems of its component parts. It summarises the book's argument and its implications for our understanding of both legal change and national versus regional historical development.*

Houston, R. A. *The Coroners of Northern Britain c. 1300–1700.* Basingstoke: Palgrave Macmillan, 2014. DOI: 10.1057/9781137381071.0005.

For the last 800 years coroners have been important features of England's legal and political landscape, best known as investigators of sudden, suspicious or unexplained death. Transplanted early on to Wales (1284) and (on paper at least) Ireland, the office exists today in a recognisable form over much of the English-speaking world. In contrast, historians have largely ignored, misunderstood or dismissed the office of coroner in Scotland. On the one hand is the axiom of English history that 'The office of Coroner is a uniquely English institution ... Scotland, of course, never had coroners' and, in his seminal work on early English coroners, historian and archivist Roy Hunnisett made almost no mention of Scotland.[1] On the other hand is a vague sense that, even if Scotland did have coroners, there is a 'lack of evidence that they were important or even necessarily active'.[2] As with many lesser royal officials across Britain and Ireland, studies that probe more prominent and enduringly successful officers, like sheriffs and sheriffs-depute, neglect the Scottish coroner. The Scottish sheriff was, from the twelfth century, 'the king's judicial, financial, administrative and military officer'.[3] Scottish historian John Pinkerton recognised the imbalance, writing as long ago as the eighteenth century. 'The duties of the sheriff have been frequently explained; but the Coroner, an office of high importance in various stages of our history, seems unknown to our legal or antiquarian enquiries.'[4] The coroner is, indeed, one of many offices of significance to late medieval and early modern Scottish history, which remain poorly understood.[5]

Medieval English coroners are far better known, although those who use the inquest records of their early modern successors study the office itself indirectly, while exploring social topics such as suicide and gender.[6] Historian Jim Sharpe only recently launched a preliminary survey into inquests in Cheshire, though again focusing on the insights they give into social history.[7] Historian Steven Gunn's important ESRC project on Tudor inquests also has this angle.[8] The present study looks at the office of coroner and its place in the different judicial, political and social systems of northern Britain. It is primarily an analysis of Scottish coroners, which draws comparisons with the distinctive history of the office in the north of England and in Wales – as well as in the south of England, which is usually (if misleadingly) held to be typical of all English regions. Thus, the first chapter outlines what coroners did in late medieval and early modern England. Chapter 2 explains how Scots looked into sudden or suspicious death, which was the post-medieval English coroner's principal function.

DOI: 10.1057/9781137381071.0005

The way Scots investigated death was certainly different from England and it is often said, in consequence, that there was and is no coroner in Scotland. That is untrue, though it seems clear that Scottish coroners never concerned themselves with sudden death unless pursuing someone who had allegedly committed a murder or 'slaughter' (killing); instead, ordinary magistrates had this investigative responsibility. The remainder of the book shows what people called 'coroners' or 'crowners' (or variant spellings thereof) did in Scotland and how their role and development compares with England. Investigating sudden death was not the only duty assigned to the English coroner, and his Scottish equivalent was also multi-functional, albeit performing different tasks, including quasi-military roles. The sometimes protean nature of the office will become clear when explaining these functions, as will the considerable breadth and slippage in the use of the title in different Scottish regions, contexts and time periods. Looking at Britain (and Ireland) as a whole, legal historians Frederick Pollock and Frederic William Maitland's description of the jury applies equally to coroners: 'We have here a plastic institution, which can assume divers shapes in ... England and Scotland'.[9]

The most obvious contrast with England was that Scottish coroners dealt more with living miscreants and their assets than with the bodies and goods of the dead; they were men of action who had a robust role in the administration of justice. There are also differences in historical origins, the status of incumbents, the means of their appointment, their relations with other judicial officers and finally in the development of the role over time. Scottish coroners functioned as executive judicial officers who serviced circuit courts, which dealt especially (if not exclusively prior to 1532) with criminal matters. The bulk of the book explains their jobs and how, over time, they became sidelined by changes in the constitution of courts and in legal procedures. We shall see that Scottish coroners were active in enforcing the pleas of the crown (murder, rape, arson and robbery) and they probably did not wholly lose these functions until the reorganisation of central criminal justice in the decades after the Restoration of the monarchy in 1660.[10] In their late-medieval heyday, their fortunes were closely tied to the justice-ayres. An ayre was 'the holding and concomitant progression of a justiciar's court along its chosen geographical route'.[11] Late-medieval justiciars were the highest officers under the crown, responsible particularly for justice ('pursuing and judging criminals, hearing appeals from the sheriff's courts and arbitrating in land cases'), but more broadly for overseeing royal government.[12]

DOI: 10.1057/9781137381071.0005

Taking of inquests in cases of sudden or suspicious death was only part of the role originally envisaged for English coroners, whose main duty was to serve the general eyre by keeping a record for presentation of the pleas of the crown (in England the greater offences were tried in the king's courts). Active between eyres and the assizes which replaced them, coroners helped the flow of justice and protected the king's financial interests.[13] With the demise of eyres after 1294 English coroners continued to serve commissions of oyer and terminer, and trialbaston (special judicial commissions to try felonies and trespasses), but their work increasingly focused on inquests, abjuration of the realm (exiling confessed felons) and attending the county court to record exactions to outlawry (summons) – sometimes as jurors.[14] The office of coroner in England became debased after c.1300, not only in status, but also in the tasks carried out. Collection of coroners' rolls became more sporadic and the rolls themselves are of widely varying quality in the fourteenth century; very few survive from the fifteenth century. After 1487, central control tightened and coroners had to file inquest verdicts with assize judges or King's Bench.[15]

Although historians have barely whispered about Scottish coroners, a fuller analysis of their origins and changing roles in justice and government allows us to speak more clearly about the nature of society and politics in different parts of the British Isles. The aim of this book is not simply to add Scottish experience to the main line of English history and so create a more comprehensive British perspective, but also to open up new questions and debates within British history, using Scotland as a starting point for comparison. Legal change in Scotland mostly happened more quickly than in England, notably the secularisation of church courts at the Reformation and the rapid development of new means of dealing with debt and credit. One example of a prolonged transition, in contrast, was the emergence of the 'Session' (later the Court of Session) in the century before the foundation of the College of Justice in 1532. Justiciars stopped judging in civil matters after 1532. A second example is the gradual displacement of ayres in criminal matters by the central Justiciary Court (a tribunal based in Edinburgh and administering justice at fixed terms) during the century after 1532.

The remit of English coroners also shrank over time, though the reasons were quite different from Scotland. According to Hunnisett, the English coroner's role changed slowly but inexorably. 'The coroner thus had no important duty taken from him in the later Middle Ages, but

DOI: 10.1057/9781137381071.0005

some of them came to exercise him far less frequently, and he, like the sheriff, became relatively less important with the rise of the escheator and the J.P.'[16] The erosion of the coroner's significance came about in Scotland thanks to new ways of prosecuting at the Justiciary Court, but also because of the *growing* power of the sheriff as the hub of civil and criminal justice in Scotland's shires. Scottish Justices of Peace (sic) were of lesser importance until the eighteenth century and, even then, had narrower remits than their English counterparts.

As with the wider development of Scotland's central courts and their procedure, there was no moment at which the crown abrogated in its entirety the older framework, that included coroners, and no tidy reform at one point in time (not even in 1672, though perhaps in 1709 as we shall see later). Instead a gradual process of substitution occurred – or perhaps the grafting of a new form onto an existing one.[17] Fifteenth-century judicial and administrative experiments ran in many directions for centuries afterwards, responding to a variety of contingencies, and manifesting multiple institutional options. The Scottish crown of the late Middle Ages was strong enough to enforce justice, but it chose to bring about change by providing alternatives rather than direct replacements, aided by the fluidity of institutional forms and by personalised under-standings of service, which meant that almost no official had a precise job description. Rulers could argue that something new was really just an older form under a different name. Ultimately, the frictions between different frameworks of justice gave way, in the century after 1650, to a single set of officials, courts and procedures – in which the Scottish coroner had no part.

The functions and development of the office of coroner in different parts of Britain (and Ireland, which merits some analysis) show that English coroners were independent judicial office holders who formed a structural part of a transparent, participative system of justice, whereas their Scottish namesakes were judicial agents. Whereas English coroners held inquests in public before a jury, Scottish magistrates who investi-gated deaths acted and decided in private, informing themselves as they saw fit. What happened to coroners therefore illuminates the very dif-ferent legal and political histories of Scotland and England. Chapter 5 nevertheless teases out some intriguing similarities in the functions of franchisal coroners in northern England and coroners as a whole in Scotland up to the sixteenth century, arguing that the north of Britain and possibly also Wales shared social and administrative characteristics,

DOI: 10.1057/9781137381071.0005

which made them different from southern and Midland England. In particular, peace-keeping in the north and west of Britain involved kin and lords, whereas in the south and east responsibility for local enforcement lay with communities. Thus the book also highlights the importance of regional experience, seen through the lens of the work of coroners, to understanding British historical development.

Notes

1   B. Knight, 'History of the Medieval English Coroner System', in http://www. britannia.com/history/coroner1.html. R. F. Hunnisett, *The medieval coroner* (Cambridge: Cambridge UP, 1961). P. Lloyd, 'The coroners of Leicestershire in the early fourteenth century', *The Leicestershire Archaeological and Historical Society Transactions* 56 (1980–1), 18–32.

2   J. Goodare, *The government of Scotland, 1560–1625* (Oxford: Oxford UP, 2004), 217n.

3   D. M. Walker, *The Scottish legal system: an introduction to the study of Scots law* (1959. 6th edition. Edinburgh: W. Green/Sweet & Maxwell, 1992), 86.

4   J. Pinkerton, *A history of Scotland from the accession of the house of Stuart to that of Mary* 2 vols (London: C. Dilly, 1797), vol. 1, 385.

5   P. J. Murray, 'The lay administrators of church lands in the fifteenth and sixteenth centuries', *Scottish Historical Review* 74, 197 (1995), 26–44. R. A. Houston, 'What did the royal almoner do in Britain and Ireland, c.1450–1700?', *English Historical Review* 125 (April 2010), 1–35.

6   M. MacDonald and T. R. Murphy, *Sleepless souls: suicide in early modern England* (Oxford: Oxford UP, 1990). G. Walker, *Crime, gender and social order in early modern England* (Cambridge: Cambridge UP, 2003).

7   J. Sharpe and J. R. Dickinson, 'Coroners' inquests in an English county, 1600–1800: a preliminary survey', *Northern History* 48 (2011), 253–69. P. J. Fisher, 'The politics of sudden death: the office and role of the coroner in England and Wales, 1726–1888' (University of Leicester Ph.D., 2007), is closer in approach to what is intended here, as is M. Lockwood, 'Death, justice and the state: the office of coroner in England, 1550–1750' (Yale University Ph.D., 2014).

8   Merton College, Oxford.

9   F. Pollock and F. W. Maitland, *The history of English law before the time of Edward I* 2 vols (1895. 2nd edition, Cambridge: Cambridge UP, 1898), vol. 2, 625.

10 P. G. B. McNeill and H. L. MacQueen (eds), *Atlas of Scottish history to 1707* (Edinburgh: Scottish Medievalists and Department of Geography, University of Edinburgh, 1996), 223.

DOI: 10.1057/9781137381071.0005

11 A. M. Godfrey, *Civil justice in Renaissance Scotland: the origins of a central court* (Leiden: Brill, 2009), 18–19.

12 G. W. S. Barrow, *The kingdom of the Scots: government, church and society from the eleventh to the fourteenth century* (London: Edward Arnold, 1973), 83. Quotation from M. Brown, *The wars of Scotland, 1214–1371* (Edinburgh: Edinburgh UP, 2004), 35.

13 A. Harding, *The law courts of medieval England* (London: George Allen & Unwin, 1973), 74.

14 Hunnisett, *Medieval coroner*, 197. H. M. Jewell, *English local administration in the Middle Ages* (Newton Abbot: David & Charles, 1972), 154. E. Powell, *Kingship, law, and society: criminal justice in the reign of Henry V* (Oxford: Oxford UP, 1989), 60, 74–5. D. J. Clayton, *The administration of the county palatine of Chester, 1442–1485* (Manchester: Manchester UP, 1990), 190–1. J. C. Cox, *Three centuries of Derbyshire annals as illustrated by the records of the Quarter Sessions* 2 vols (London: Bemrose, 1890), vol. 1, 67–8. C. Burt, 'The demise of the general eyre in the reign of Edward I', *English Historical Review* 120 (2005), 1–14.

15 3 Hen. VII c. 2.

16 Hunnisett, *Medieval coroner*, 199. An escheator was a gentleman appointed for a year at a time by the Lord Treasurer to collect payments for wardships and other feudal dues, and to claim forfeitures. Escheators were one of the offices created in the thirteenth century to increase central control over justice, administration and finance.

17 Godfrey, *Civil justice*, 451–2.

DOI: 10.1057/9781137381071.0005

1

Coroners in England, Wales and Ireland: An Overview of the Development of Their Roles

Abstract: *English coroners originated in the twelfth century. This chapter summarises the subsequent development of their different legal and administrative roles over time in England, Wales and Ireland, charting how a figure with high social status and broad judicial functions, who helped the flow of justice and protected the king's financial interests, gradually became confined to presiding over inquests, before juries, into sudden, suspicious or unexplained deaths. The development of a peripatetic royal judiciary holding assize courts and an active county magistracy (Justices of the Peace) made the power of coroners over anything except determining causes of death largely vestigial by the fifteenth century; it also diminished their social standing. The chapter gives examples of how deaths were reported, how inquests operated and how coroners were appointed and remunerated, mostly in England.*

Houston, R. A. *The Coroners of Northern Britain c. 1300–1700.* Basingstoke: Palgrave Macmillan, 2014. DOI: 10.1057/9781137381071.0006.

'[L]ong regarded as a quiet and curious backwater of the English legal system', the workings of coroners' inquests and the duties of the coroner are nevertheless quite well understood.[1] From 1194 the English crown charged coroners with investigating sudden or suspicious deaths, to determine whether the cause was natural, accidental or through human agency.[2] English coroners were royal officials who sat with a jury 'on the view of' a corpse (they had to see it); inquests were held before them rather than by coroners. A dead body subject to a coroner's inquest required his warrant before it could legally be buried. More broadly, coroners tackled serious crime and protected royal financial rights; they recorded their work in a separate roll of the pleas of the crown ('Corone' – hence Coroner) as a check on the sheriff.[3] All private criminal accusations (appeals) made in the county court had to go through a coroner as part of the procedure; the coroner supervised sanctuary and abjuration of felons until 1623; he took evidence from felons (turned 'approvers') against other felons; he had to attend gaol delivery sessions; he had the power to bind over suspects and witnesses to appear at general eyres or, from the fourteenth century, assizes; he could commit to prison or to trial; he secured the chattels of convicted felons.[4]

This last point is often ignored, yet it remained potentially an important expectation of coroners until the abolition of forfeiture for felony in England in 1872.[5] 'By the common law, after a felon be found guilty before the Coroner ... there the Coroner, Sheriff, undersheriff, or Escheator, etc. may (for the King) seise the goods of the felon, and praise them by an Inquest, etc. ... for by such thing found before the Coroner, the goods of the Felon are forfeited without further inquiry ... and yet the Officer may not in such case carry the Felons goods away, but ... must leave them in the custody of the Felons Neighbours ... to be answered to the King'.[6] Though the court was one of inquiry rather than prosecution, an appropriate finding was itself sufficient authority for criminal proceedings; the coroner could return indictments based on his own inquest or coming from a jury of presentment.[7] Elizabethan commentator Sir Thomas Smyth described the return as 'in the nature of an indictment, which is not a full condemnation'.[8]

Late-seventeenth-century writers on judicial practice, such as William Nelson, noted that officers generally secured goods after the arrest of a living person on suspicion of felony, partly to prevent their disposal and partly to ensure the accused had material support while awaiting trial – which might take place days, weeks or even months later.[9] Until

DOI: 10.1057/9781137381071.0006

conviction, an accused felon could dispose of assets as they wished, but fraudulent transfer to frustrate the crown was illegal. In the words of the Privy Council, writing to Sheriff Sir Nicholas Bacon in 1604, a seizure should occur 'in such sort as there may be no fraud committed by embezzling, or by lawful conveyance of' the assets in question.[10]

In most other regards coroners' remits shrank from c.1300 until, from Tudor times, they mainly investigated sudden or suspicious death. By that stage, constitutional writers could rationalise coroners quite differently from their initial conception. For Sir Thomas Smyth, writing in 1583, 'this name commeth because that the death of everie subject by violence is accounted to touch the crowne of the Prince, and to be a detriment to it, the Prince accounting that his strength, power, and crowne doth stande and consist in the force of his people, and the maintenance of them in securitie and peace'.[11] Elizabethan and Jacobean legal authorities too dealt only with the investigation of death.[12] Had they fulfilled their original purpose, English coroners would have superseded (for example) court leet juries entirely in preliminary enquiries into felonies and also perhaps taken over the presentment of felons to the general eyres.[13] Henry II's reforms had removed the pleas of the crown from the sheriffs and placed them exclusively with justices in the king's household or commissioned by the king, the coroner recording accusations and pleadings between eyres so that trials could proceed swiftly when the judges arrived. The focus on dead bodies came out of Norman interest in the murder of fellow countrymen and because homicide was an important source of forfeiture, the coroner listing and thus asserting and protecting the king's rights.[14] Magna Carta confirmed the reduced power of English sheriffs and a statute of 1461 (1 Ed. IV, c. 2) forced them to notify a felony to Justices of the Peace (JPs) rather than to move against the suspect themselves; in criminal matters coroners' rolls became more authoritative as a record in the thirteenth century and the coroner had already superseded the sheriff in important judicial and administrative roles by 1307.[15] Between the twelfth and the fourteenth centuries, the English sheriff 'evolved from a regional dictator with true executive authority into a tightly regulated bureaucrat whose chief administrative purpose was to respond to a multiplicity of royal writs'.[16]

English sheriffs had largely ceased to be hereditary in the twelfth century.[17] For their part, late medieval and later coroners were normally appointed for life. Over time, the rise of the Commission of the Peace meant that JPs, who were usually of higher status than coroners, came to have greater responsibility for the peace-keeping aspects of coroners' work.[18] The same practical supersession explains the earlier decline of serjeants of the peace, 'a body of officers, both royal and baronial, specially entrusted with the preservation of peace, the repression of crime, and the execution of the orders of the courts of justice. Those duties included making arrests and distraints, the placing of persons under attachment by sureties to appear for trial, the services of summonses, the carrying of official messages, and the collection of some of the profits of jurisdiction.'[19] At the same time, the demise of English justices in eyre during the thirteenth century, gone by 1265 to be replaced by justices of the Benches, cemented the lesser role.[20] 'The office of coroner was at its zenith in the second half of the thirteenth century.'[21] Together, a peripatetic royal judiciary and an active county magistracy made the power of coroners over anything except sudden death largely vestigial by the fifteenth century, except (as will become clear) in the north of England.[22] Where coroners had once kept the pleas of the crown across the realm, commissioners of the peace came to hold and determine them. A decrease in appeals of felony, and abolition of outlawry and of the *murdrum* fine (1340; this was a sanction against a subdivision of a shire called a 'hundred', in cases where the killer of a murdered man could not be identified) further rendered aspects of their existing duties obsolete by the end of the Middle Ages.

Alongside contracting functions went diminishing status. Writing in the mid-eighteenth century, jurist William Blackstone believed that an early English coroner had to be a knight with an estate 'sufficient to maintain the dignity of his office, and answer any fines that may be set upon him for his misbehaviour'. As originally conceived in 1194 each shire had four coroners, of whom three had to be knights and one a clerk, but knighthood was no longer a qualification by the end of the fourteenth century. Thus Blackstone, following the judge Matthew Hale and Middlesex coroner Edward Umfreville (and legislation of 1732 establishing a property qualification for English and Welsh JPs), opined that 'through the culpable neglect of gentlemen of property, this office has been suffered to fall into disrepute, and get into low and indigent hands' of men who only wanted remuneration.[23] Those who followed Blackstone in condemning the debasement of the office included county historian Edward Hasted, who took a very 'county' line when he wrote in the 1780s: 'this office has been suffered to fall into the hands of those of lower rank, being at present usually executed, in this county in particular [Kent], by attornies at law.'[24]

DOI: 10.1057/9781137381071.0006

This social change had happened no later than the mid-sixteenth century, when lawyer and printer John Rastell observed: 'CRowner is an auncient officer of trust and of great aucthoritie, ordayned to be a principal conseruator, or kéeper of the peace, to beare recorde of the pleas of the Crowne, and of his owne sight, and of diuers other things many in number, &c. But at this day, either ye aucthoritie of the Coroner is not so great, as in fore tyme it was, whereby the office is not had in like estimation; Or els ... meane menne and vndiscréete nowe of late are commonly chosen to the office.'[25] Sir Thomas Smyth described the coroner as 'one chosen by the Prince of the meaner sort of gentlemen, and for the most part a man seene [versed] in the lawes of the Realme'.[26] Early modern English coroners had some status in their own right: Sharpe calls those of the seventeenth century 'legally educated minor country gentry', though the only formal qualification was being an independent freeholder.[27] In the late seventeenth and eighteenth centuries most styled themselves 'gentleman', but they were usually what historians class as 'pseudo-gentry' (professional men) rather than true members of the landed classes and their authority came from their learning and from bearing the king's writ (antiquary and lawyer William Lambarde described the title as one of the 'names of dignitie by reason of office onely') rather than from their personal social status.[28] Some eighteenth- and nineteenth-century incumbents seem to have used the office as a way of enhancing their social and professional standing.[29] By then (and ever since) most English coroners were men of law; for example, 83 per cent of those of known qualification in 1891 were solicitors.[30]

Blackstone appeared just after a major change in fee structure. English coroners were initially unpaid, only remunerated from local rates for each murder inquest from 1487: taking any further payment was 'plaine extorcyon'.[31] English coroners acquired a wider remit over (and a closer focus on) dead bodies under the early Tudors and were, in 1510 (1 Hen. VIII, c. 7), exhorted to hold inquests into those 'slayne by myssaventre' and, from 1554, had to keep records of their examination of accused criminals.[32] The 1510 statute tried to deal with resistance to holding inquests on bodies for which they would receive no fee (none was payable for accidental death) and was part of a series of initiatives in the 1510s and 1520s to push government into the localities.[33] In reality, however, boroughs routinely paid their coroners for all sorts of inquest. Sometime in 1593 or 1594, for example, the town of Sheffield paid its coroner 7/- 'for his fee or paynes for ye death of yt maid who fellonyouslie

DOI: 10.1057/9781137381071.0006

kylled herself by cuttinge of her owne throat at Jo. Crewswickes'.[34] Only in 1752 (25 Geo. II, c. 29) did an act arrange for county coroners to be paid for all inquests provided their jurisdiction paid local rates according to 12 Geo. II, c. 29.[35]

The office of coroner has survived for 800 years by evolving to meet the changing needs of government and society. Until 1888, county freeholders elected most English crown or 'general' coroners at the county court in a country with a stronger tradition of both centralised control and participative democracy than Scotland. Corporations, colleges or lords of liberties with appropriate charter privileges appointed franchisal or 'special' ones, though they were still royal officers with the same relationship to the justices. Newcastle-upon-Tyne is an example of a town with this privilege.[36] Franchisees appointed a fifth of all English coroners from the thirteenth to the late nineteenth century.[37] Because most English coroners were elected locally, they were responsive to local social and political change, dealing with issues which themselves became politicised and thus subject to political intervention.

One example is the treatment of 'deodands' (animate or inanimate objects that independently caused the death of a person). Identified and valued by coroners' juries so they could be forfeited or compounded for with the crown, parliament abolished deodands in 1846 because, for a century, inquests had been able to turn a criminal into a delictual liability, enabling them to use deodands as a way of penalising what they construed as corporate negligence.[38] That English coroners' inquests could handle deodands in this way shows their flexibility and also their ability to consider extraneous matters such as civil and criminal liability; it also shows how very different their inquests were from any other English court. Coroners had guidelines on how an inquest should be conducted and how to frame the verdict, but they had no clear legal definition to their purpose.[39] Furthermore, the standard of proof required before an inquest was so persistently low as to allow any interpretation of liability, not only with deodands, but also with distinguishing between culpable suicides (*felones de se*) and blameless ones (*non competes mentis*).

In the early modern period and ever since, some commentators described inquests as at best variable in functioning, at worst idiosyncratic or even corrupt. English coroners could in theory be removed or have their verdicts overturned or 'traversed', but for most purposes they were responsible to no higher authority in anything expect procedural matters (see below). To this day, coroners' inquests are not bound by the normal 'rules of evidence' used by the English and Welsh constabulary and all other criminal courts in these parts of the United Kingdom; this accounts for the removal in 1977 of their power to assign 'guilt' and to commit to trial, and for other restrictions since placed on them.[40]

Unable to initiate investigations, English coroners relied on information from individuals with knowledge of a sudden or suspicious death. Notification came from lay members of the public and amateur officials such as parish constables; in the eighteenth and early nineteenth century from medical men and police constabulary; and ultimately, from 1837, registrars. Coroners issued a warrant to local constables to list potential jurors and the juries they used varied in size from 10 to 24 men, usually all neighbours of the deceased. The inquest was a local and participative event, held in a private house or public building such as a tavern or church porch in the vicinity where the body was found, the corpse's location alone determining jurisdiction for a coroner's inquest, which absolutely had to view the body. A sense for how public coroners' proceedings could be comes from the diary of Yorkshire vicar Robert Meeke. On 6 October 1692, Meeke went to see 'a woman who had cut her throat. Endeavoured to convince her of the greatness of her sin, and the danger she was in, the mercy of God had prevented her, that she could not do as she wanted.' Ten days later he recorded: 'Many went to Marsden, to hear and see what was said and done by the coroner and jury, who were met to examine persons concerning the aforesaid woman's action, who cut her throat, and is now dead; and ordered to be buryed at a lane end in Lingarths.'[41]

Other procedures varied considerably between individual coroners, the system reflecting 'the prejudices, habits, and values of each particular locality in all the diversity of its public and private interests, conflicts, and routines'.[42] Inquests seldom sought evidence from medical practitioners prior to the nineteenth century. Surgeon Thomas Brugis's mid-seventeenth-century *Vade mecum or, a companion for a chyrugion* offers an outline of 'the manner of making Reports to a Magistrate, or Coroner's Inquest', but the account merely summarises a passage from Ambroise Paré's guidance on presenting before a judge in France and is better as a guide to Roman than common law practice.[43] Medical men rarely appeared as witnesses before northern English coroners' inquests into suicide and, when they did, talked about physiological aspects of wounds and ingestions rather than psychological dimensions of their patients. Inquests sat on the view of a body and the evidence that interested them lay principally before their eyes; witnesses of any kind simply helped the jury to see more clearly.

This is not to say that English coroners' inquests had no supervision, because procedural regulation existed at common law. English coroners were always closely accountable to the crown through a centrally located or centrally controlled judiciary and the government's hand became even firmer under the Tudors.[44] During the sixteenth century, both attorney general and Privy Council can be found intervening to ensure that inquests 'take speciall care to the due proceeding'.[45] The records of county coroners' inquests usually survive among the papers of courts that supervised them, with coroners subject to routine monitoring by justices and by the Lord Chief Justice of the King's Bench (as principal coroner).[46] English common law required conformity to set forms and overturning an inquest verdict required evidence of failure of procedure rather than substance. JPs could hold inquests themselves if no corpse was available, as could assize judges, both without a commission.[47] JPs paid coroners' expenses, meaning that supervision remained close even after an act of 1693 (4 & 5 W. & M. c. 22) removed the need for coroners to file inquest findings with King's Bench. Subject to this oversight, an English coroner could be removed from office for 'extortion, wilful neglect of duty, or misdemeanour in his office' and some, pursued before Star Chamber, King's Bench, Chancery, Exchequer or even assizes, suffered dismissal.[48]

The other central official who could involve himself in certain cases, such as deodands and the forfeiture of the goods of felons of themselves (suicides), was the royal almoner. A Jacobean Star Chamber case illustrates his interest and the role of others.[49] Michael Barker of Corrybecke in Cumberland drowned on 5 October 1621. Relying on reported public opinion, the almoner characterised the death as a *felo de se* (deliberate and thus felonious self-murder). The almoner dismissed the jury, allegedly directed by the defendants John Jenkinson and Jane Barker (the widow) to find that Michael died by accidental drowning, as 'being all or for the most parte men of very simple and weak understanding and judgement'. The almoner went on to insist that general opinion held it as wilful death and that Jenkinson and the widow had used devious and improper means to influence the verdict. Yet as with all the suits where both complaints and answers survive, the apparent confrontation is misleading. In his answer, Jenkinson acknowledged that the jury had found 'against the king', but complained that the deputy almoner, Cuthbert

DOI: 10.1057/9781137381071.0006

Orfure or Orfeur, had bullied them to change their verdict and the coroner, Richard Robinson, to refuse to register it. Jenkinson and the foreman of the jury claimed to have felt so intimidated by Orfure's attempts to 'terrify and threaten' them that they asked Sir Thomas Chamberlin, an assize judge, to investigate the matter. The defendant reported that Chamberlin had ordered the coroner to accept the verdict, saying he had 'done the jury wrong' to refuse to do so. Jenkinson further claimed that he did not know the deceased or his wife until, three months after his death, he met the widow, Jane, and subsequently married her. Occurring two years after the suicide and inquest, the Star Chamber suit was a continuation of a local power struggle in a central court. Patronage networks in early modern England were fluid and multi-valent and those pressed by one royal officer (an almoner) thought nothing of using another (an assize judge) against him.

Coroners in Tudor and later Wales did the same job as in England. Formally instituted in parts of Wales during the thirteenth century, they may have been active in some places from the early fifteenth century, but only became functional as intended after the shiring of the whole principality in Henry VIII's reign.[50] Elected in the county court, coroners can be found in Ireland from the thirteenth century, notably in Dublin, but they were not widespread, even in the sixteenth century. They probably adjusted their roles to local circumstances (in the sixteenth century there were still serjeants of the peace in Ireland, an office with British origins discussed later) and at no time did coroners form part of the unitary legal system that Angevin kings had tried to impose.[51] There was overlap and inter-changeability between coroners, sheriffs and justices. More, the nature of social organisation was different, close in some ways to the personal bonds that sustained society and order in Scotland and the north of England, discussed in Chapter 3. As historian Steven Ellis puts it: 'Gaelic lordship was more a lordship of men than land.'[52] Arguably also the whole idea of 'felony' was subordinated to that of tort.[53] The sense one gets from patchy documentation is that coroners were geographically restricted in Ireland; most sixteenth-century examples of their actual operations (as distinct from their provision on paper) come from Dublin (where annually elected bailiffs executed the office) and look like responses to short-lived initiatives.[54] Despite the revival of assize courts and quarter sessions in the early seventeenth century, there is little sign that Irish coroners were either widespread or active, even within the Pale, the 'Elizabethan' counties and Anglo-Irish lordships.[55] In

DOI: 10.1057/9781137381071.0006

the eighteenth century also they operated mostly in the towns, though there is some evidence of regular appointments elsewhere, especially in the north and east, and of coroners functioning as they did in England and Wales.[56] Even more than Wales, Ireland remained an unstable and fragmented polity until late in the early modern period, where Scotland was settled and centrally governed from a much earlier date. There is no firm evidence that coroners operated across Ireland as their English masters intended, prior to extensive reorganisation in the mid-nineteenth century.[57]

Notes

1 J. Sim and T. Ward, 'The magistrate of the poor? Coroners and deaths in custody in nineteenth-century England', in M. Clark and C. Crawford (eds), *Legal medicine in history* (Cambridge: Cambridge UP, 1994), 245.

2 G. Gross (ed.), *Select cases from the coroners' rolls A.D. 1265–1413 with a brief account of the office of coroner* (London: Selden Society, 1896), xiii–xxxvi. A. T. Carter, *A history of the English courts* (1899. London: Butterworth & Co., 1927), 118–21. J. D. J. Havard, *The detection of secret homicide* (London: Macmillan, 1960).

3 R. F. Hunnisett, 'The origins of the office of coroner', *Transactions of the Royal Historical Society* 5th series 8 (1958), 85–104. A. L. Brown, *The governance of late medieval England, 1272–1461* (London: Edward Arnold, 1989), 111–12. C. Smith, 'Medieval coroners' rolls: legal fiction or historical fact?', in D. E. S. Dunn (ed.), *Courts, counties and the capital in the later Middle Ages* (Stroud: Sutton, 1996), 95. J. G. Bellamy, *Crime and public order in England in the later Middle Ages* (London: Routledge & Kegan Paul, 1973), 108, 112. J. Hudson, *The Oxford history of the laws of England. Volume II, 871–1216* (Oxford: Oxford UP, 2012), 507–9.

4 G. Jacob, *The law-dictionary ... enlarged and improved by T. E. Tomlins*, 2 vols (1729. London: Andrew Strahan, 1797), 'coroner'. Gross (ed.), *Select cases*, xxiv–xxx. Jewell, *English local administration*, 155. J. H. Langbein, *Prosecuting crime in the Renaissance: England, Germany, France* (Cambridge, MA: Harvard UP, 1974), 14–15.

5 33 & 34 Vict. c. 23.

6 R. Bolton, *A Justice of Peace for Ireland* (1638. Dublin: Benjamin Tooke and John Crooke, 1683), book 1, 88. 1 R. III c. 3.

7 T. Smith, *The parish. Its obligations and powers: its officers and their duties* (London: S. Sweet, 1854), 333–7. Powell, *Kingship, law, and society*, 66–7. Langbein, *Prosecuting crime*, 13–15.

DOI: 10.1057/9781137381071.0006

8   T. Smyth, *De republica Anglorum. The maner of Gouernement or policie of the Realme of England* (London: H. Midleton, 1583), 2:21 (p. 73).

9   W. Nelson, *The office and authority of a Justice of Peace* ... (1704. 3rd edition. London: John Nutt, 1710), 301.

10   University of Chicago, Regenstein Library, Bacon Mss., 4180, Privy Council to Sir Nicholas Bacon, 17 September 1604. Others doubted shrieval probity. In 1575, Thomas Mydlemore presented a petition to Queen Elizabeth arguing that sheriffs had hidden forfeitures and profits in their county courts. NA SP 12/106/67.

11   Smyth, *De republica Anglorum*, 2:21 (p. 73).

12   J. Wilkinson, *A treatise collected out of the statutes of this kingdom* ... *concerning the office and authoritie of coroners and sherifes* (London: Iohn Wilkinson, 1618), 1–49.

13   F. J. C. Hearnshaw, *Leet jurisdiction in England, especially as illustrated by the records of the court leet of Southampton* (Southampton: Cox & Sharland, 1908), 95.

14   A. Harding, *A social history of English law* (Harmondsworth: Penguin, 1966), 49.

15   W. A. Morris, *The medieval English sheriff to 1300* (Manchester: Manchester UP, 1927), 238–9.

16   R. Gorski, *The fourteenth-century sheriff: English local administration in the late Middle Ages* (Woodbridge: Boydell, 2003), 2.

17   Lancashire and Westmorland were among the last hereditary sheriffs to go. Morris, *Medieval English sheriff*, 179. H. M. Cam, 'Shire officials: coroners, constables, and bailiffs', in J. F. Willard, W. A. Morris and W. H. Dunham (eds), *The English government at work, 1327–1336. Volume III, local administration and justice* (Cambridge, MA: Medieval Academy of America, 1950), 153.

18   Brown, *Governance of late medieval England*, 112. Hunnisett, 'Origins', 90. Hunnisett, *Medieval coroner*, 191–9.

19   R. Stewart-Brown, *The serjeants of the peace in medieval England and Wales* (Manchester: Manchester UP, 1936), 73. For a satirical description of their work from the early Stuart period, see J. Earle, *Micro-cosmographie, or, A peece of the world discovered in essayes and characters* (1628. 6th edition, 1633, reprinted. London: E.C., 1903), 121–2.

20   D. M. Walker, *A legal history of Scotland* 7 vols (Edinburgh: W. Green etc., 1988–2004), vol. 1, 216.

21   Hunnisett, *Medieval coroner*, 190.

22   Ibid., 198–9. Clayton, *Administration of Chester*, 190.

23   5 Geo. II, c. 18. W. Blackstone, *Commentaries of the laws of England* 4 vols (Oxford: Clarendon Press, 1765–9), I.9.II [vol. 1, 336]. M. Hale, *History of the pleas of the crown* 2 vols (London: E. and R. Nutt, and R. Gosling, 1736), vol. 2, 222. E. Umfreville, *Lex Coronatoria* (London: R. Griffiths; and T. Becket, 1761), v. Gross (ed.), *Select cases*, xx. Cam, 'Shire officials', 150, 152. Lloyd, 'Coroners of Leicestershire', 20–8.

24 E. Hasted, *The history and topographical survey of the county of Kent* 12 vols (2nd edition. Canterbury, 1797–1801), vol. 1, 214–-15.

25 J. Rastell, *An exposition of certaine difficult and obscure words, and termes of the lawes of this realme* (London: R. Totteli, 1579), 51–2. Rastell died in 1536 and his work was first published in 1563.

26 Smyth, *De republica Anglorum*, 2:21 (p. 72).

27 J. A. Sharpe, *Crime in seventeenth-century England: a county study* (Cambridge: Cambridge UP, 1983), 34.

28 W. Lambarde, *Eirenarcha* (London: [Adam Islip], 1610), 489.

29 D. Eastwood, *Governing rural England: tradition and transformation in local government, 1780–1840* (Oxford: Oxford UP, 1994), 68.

30 F. W. Lowndes, *Reasons why the office of coroner should be held by a member of the medical profession* (London: J. & A. Churchill, 1892), 32. Physicians or surgeons appointed as coroners had to give up private practice, but lawyers did not. H. R. Bickerton and R. M. B. Mackenna, *A medical history of Liverpool from the earliest days to the year 1820* (London: J. Murray, 1936), 148.

31 Huntington Library 24872: [Fitzherbert] *In this booke is conteyned the offices of Shyriffes...* (1556) [np]. 1 Hen. VIII, c. 7, prohibited the taking of fees for death by misadventure. Gross (ed.), *Select cases*, xxi.

32 R. F. Hunnisett (ed.), *Calendar of Nottinghamshire coroners' inquests, 1485–1558* (Nottingham: Thoroton Society, 1969), xviii. A. Murray, *Suicide in the middle ages* 2 vols (Oxford: Oxford UP, 1998, 2000), vol. 1, 132–9. Harding, *English law*, 127. 1 and 2 Philip and Mary, c. 13.

33 P. Slack, *From reformation to improvement: public welfare in early modern England* (Oxford: Oxford UP, 1999), 15–16. The 1510 act focused on deaths requiring subsequent judicial procedure. [Brodrick] 'Report of the committee on death certification and coroners...1971', PP XXI (1971–2), [Cmnd. 4810], 112.

34 J. D. Leader (ed.), *Extracts from the earliest book of accounts belonging to the town trustees of Sheffield...1566–1707* (Sheffield: Leader and Sons, 1879), 68.

35 25 Geo. II, c. 29, restricted English coroners primarily to investigating deaths.

36 Jacob, *Law-dictionary*, 'coroner'. Cox, *Derbyshire annals*, vol. 1, 66–95. Cam, 'Shire officials', 157.

37 'Brodrick Report', 107–8. In 1868, the figure was 64 of 334 (19 per cent). Fisher, 'Sudden death', 221.

38 Blackstone, *Commentaries*, I.8.XVI [vol. 1, 291]. W. Nelson, *Lex maneriorum, or, the law and custom of England relating to manors and lords of manors, their stewards, deputies, tenants and others: viz. of the lords right to deodands, felons goods, waifs, strays, wrecks, and goods of felo de se...* (London: E. & R. Nutt and R. Gosling, 1726), 72–4. R. H. Wellington, *The king's coroner* (London: W. Clowes, 1905), 14–18. E. Cawthon, 'New life for the deodand: coroners' inquests and occupational deaths in England, 1830–46', *The American*

DOI: 10.1057/9781137381071.0006

Journal of Legal History 33 (1989), 137--47. T. Sutton, 'The deodand and responsibility for death', *Journal of Legal History* 18 (1997), 44–55. T. Sutton, 'The nature of the early law of deodand', *Cambrian Law Review* 30 (1999), 9–20. A. Pervukhin, 'Deodands: a study in the creation of common law rules', *American Journal of Legal History* 47 (2005), 237–56. G. MacCormack, 'On thing-liability (*Sachhaftung*) in early law', *Irish Jurist* 19 (1984), 322–49.

39   Fisher, 'Sudden death', 212.

40   *The laws of Scotland. Stair memorial encyclopaedia* vol. 17 (1989), no. 975.

41   H. J. Morehouse (ed.), *Extracts from the diary of the Rev. Robert Meeke, minister of ... Slaithwaite, near Huddersfield* (London: H. G. Bohn, 1874), 57.

42   Nelson, *Lex maneriorum*, 80. O. Anderson, *Suicide in Victorian and Edwardian England* (Oxford: Clarendon Press, 1987), 40. S. O. Addy, *Church and manor: a study in English economic history* (London: George Allen, 1913), 186, 205.

43   (1651. 6th edition. London: T. Flesher, 1679), 288–95.

44   A. Alison, *Remarks on the administration of criminal justice in Scotland, and the changes proposed to be introduced into it* (Edinburgh: W. Blackwood, 1825), 10–11. 3 Hen. VII, c. 2. 22 Hen. VIII c.9.

45   J. R. Dasent (ed.), *Acts of the Privy Council of England. New series 23 A.D. 1592* (London: HMSO, 1901), 289–90, 323, 382. J. A. Guy, *The cardinal's court: the impact of Thomas Wolsey in Star Chamber* (Totowa, NJ: Rowman & Littlefield, 1977), 18, 32–3, 63, and footnotes. H. Garrett-Goodyear, 'The Tudor revival of *quo warranto* and local contributions to state building', in M. S. Arnold, T. A. Green, S. A. Sully and S. D. White (eds), *On the laws and customs of England* (Chapel Hill: University of North Carolina Press, 1981), 257–8. S. J. Gunn, *Early Tudor government, 1485–1558* (Basingstoke: Macmillan, 1995), 99. For an example from 1517 see NA STAC 2/31/53.

46   Jacob, *Law-dictionary*, 'coroner'.

47   Poph. 209; 2 Lev. 141, 152. M. Dalton, *The countrey justice ...* (1618. 2nd edition. London: A. Islip, 1626), 235.

48   Jacob, *Law-dictionary*, 'coroner'. Cam, 'Shire officials', 150–2.

49   NA STAC 8/3/16.

50   W. Rees, *South Wales and the march, 1284–1415* (London: Oxford UP, 1924), 58–9, 89. The 1284 'Statute of Rhudlan' set out regulations and structures that included coroners to govern the newly conquered principality. H. Ellis (ed.), *Registrum vulgariter nuncupatum 'The record of Caernarvon'* (London: Record Commission, 1838), 119–24.

51   G. J. Hand, *English law in Ireland, 1290–1324* (Cambridge: Cambridge UP, 1967), 60, 108–9. S. G. Ellis, *Reform and revival: English government in Ireland, 1470–1534* (Woodbridge: Boydell, 1986), 201–4. A. J. Otway-Ruthven, 'Anglo-Irish shire government in the thirteenth century', in P. Crooks (ed.), *Government, war and society in medieval Ireland* (Dublin: Four Courts, 2008), 138–9. T. Skyrme, *History of the justices of the peace* 3 vols (Chichester: Rose,

1991), vol. 3, 29–42. B. Farrell, *Coroners: practice and procedure* (Dublin: Round Hall, 2000), 1–45.

52 S. G. Ellis, 'Commentary from a British perspective', in P. Blickle (ed.), *Resistance, representation, and community* (Oxford: Oxford UP, 1997), 61.

53 *Ibid.*, 62. Parts of the Tudor north of England did not operate forfeiture for felony. *Ibid.*, 58–9.

54 Farrell, *Coroners*, 12–20. J. J. Webb, *Municipal government in Ireland: mediaeval & modern* (Dublin: Talbot, 1918), 9. C. Brady, 'Court, castle and country: the framework of government in Tudor Ireland', in C. Brady and R. Gillespie (eds), *Natives and newcomers: essays on the making of Irish colonial society, 1534–1641* (Dublin: Irish Academic Press, 1986), 26. C. Tait, *Death, burial and commemoration in Ireland, 1550–1650* (Basingstoke: Macmillan, 2002), 39, 174. M. Potter, *The government and the people of Limerick: the history of Limerick Corporation/City Council 1197–2006* (Limerick: Limerick City Council, 2006), 63. C. Maginn, 'Elizabethan Cavan: the institutions of Tudor government in an Irish county', in B. Scott (ed.), *Culture and society in early modern Breifne/Cavan* (Dublin: Four Courts Press, 2009), 69–84.

55 D. Dickson, *Old world colony: Cork and South Munster, 1630–1830* (Cork: Cork UP, 2005), 15–16.

56 B. Henry, *Dublin hanged: crime, law enforcement and punishment in late eighteenth-century Dublin* (Dublin: Irish Academic Press, 1994), 38. N. Garnham, *The courts, crime and the criminal law in Ireland, 1692–1760* (Dublin: Irish Academic Press, 1996), 97–8. Garnham relies partly on Matthew Dutton's text of 1721, which is more prescriptive than descriptive. M. Dutton, *The office and authority of Sheriffs, Under-Sheriffs, Deputies, County-Clerks and Coroners in Ireland* (Dublin: S. Powell, 1721). D. A. Fleming, *Politics and provincial people: Sligo and Limerick, 1691–1761* (Manchester: Manchester UP, 2010), 127, also uses a legal text when summarising functions.

57 W. G. Huband, *A practical treatise on the law relating to the grand jury in criminal cases, the coroner's jury and the petty jury in Ireland* (London: Stevens & Sons, 1896). J. L. Leckey and D. Greer, *Coroners' law and practice in northern Ireland* (Belfast: SLS Legal Publications, 1998), 1–14. M. J. Clark, 'General practice and coroners' practice: medico-legal work and the Irish medical profession, c.1830–c.1890', in C. Cox and M. Luddy (eds), *Cultures of care in Irish medical history, 1750–1970* (Basingstoke: Macmillan, 2010), 37–56. M. McGoff-McCann, *Melancholy madness: a coroner's casebook* (Cork: Mercier, 2003).

DOI: 10.1057/9781137381071.0006

2
Investigating Sudden Death in Scotland: The Task of Local Magistrates

Abstract: *Before 1707, Scotland was a wholly separate country from England and its legal system was quite different. When it came to sudden, suspicious or unexplained death in Scotland, there was no inquest by a coroner's jury, as there was in England and Wales. Instead, local magistrates conducted inquiries in private, taking evidence from anyone who might have knowledge of the circumstances of a death, including forensic evidence from a medical examiner. Their main purpose was identifying whether the death was wrongful (i.e. the body was the victim of foul play) and thus whether a criminal prosecution was required. The chapter charts the development of more-or-less standardised Scottish investigative procedures between the sixteenth and nineteenth centuries. It also explores the reasons behind the emphasis on holding these in private, and the findings they produced. It uses mainly examples of formal legal inquiries into cases of murder and suicide.*

Houston, R. A. *The Coroners of Northern Britain c. 1300–1700*. Basingstoke: Palgrave Macmillan, 2014.
DOI: 10.1057/9781137381071.0007.

In scholarly and public minds alike, coroners and the investigation of suspicious or sudden death are closely interwoven. Some contributors to Victorian debates on the reform of investigations into sudden death in Scotland even borrowed the word 'coroner' to describe the official who performed the task, though this was by no means his title.[1] People died under dubious circumstances in Scotland, as they did anywhere, and there were plainly mechanisms for investigating causes and agencies. This chapter sets out who, when, why and how, charting the development of more-or-less standardised procedures between the sixteenth and nineteenth centuries. In short, ordinary magistrates investigated sudden deaths in Scotland, following a process called 'precognition'. A lengthy and rigorous procedure of which examples will be offered later, precognition was and is a preliminary investigation into any crime, including a death that looked as though it might be suspicious. Because this book came out of research on early modern suicide, most of the examples that follow relate to investigations into this means of dying.[2]

One example of an investigation, which highlights the differences between Scotland and England, comes from near Edinburgh in the later Stuart era. In the winter of 1687–8 the owner of a textile works in East Lothian, Sir James Standsfield, was found dead on his estate, face down in a pool of water. At first sight, it looked like suicide, Standsfield allegedly a chronic melancholic who had recently showed signs of acute depression. Those who knew the family, however, quickly became suspicious about his son Philip's role in the death, for Philip was strangely reluctant to sanction the 'sighting' or inspection of the body by lay or professional observers. At his subsequent trial for murder, the prosecution alleged,

> The pannal [pannel, or accused] did refuse to send for a chyrurgion, and to let his fathers body be sighted, though the minister, and others did expressly demand it; and the *English men* in the *Manufactury*, who were acquainted with the *Crowner-Laws*, they made a mutiny anent the burial, till the corps were sighted, yet the pannal caused bury the corps that same night without shewing them.[3]

Standsfield employed skilled immigrant workers, Englishmen living in Scotland, who in this case appealed to their own laws about the coroner or 'crowner'.

Indignant as they no doubt were, the Englishmen must have known that Scotland had no 'crowner-law'. Native Scots nevertheless expected something similar from their legal system – a public but not necessarily

DOI: 10.1057/9781137381071.0007

participative involvement – when doubt existed about the cause of death: exactly the sort of 'sighting' demanded by those suspicious of Philip Standsfield. That usually meant laying the body out in an open place like a church so that any interested person could assess it for themselves, as also happened to identify unknown bodies. Edinburgh's Tron (a market-place and weighing point) and, from the mid-seventeenth century, the Tron Kirk were examples of recognised places for this within that royal burgh, but any public building or place might suffice.

Scots also expected some sort of medical investigation by a physician or surgeon, where necessary, and this seems to have been the norm from an early date.[4] In the case of Sir James Standsfield, the Justiciary Court in Edinburgh ordered his body to be exhumed and returned to Morum church for inspection and dissection by two surgeons, their report subse-quently reviewed by the Royal College of Physicians of Edinburgh. Thus, Scots had expectations and provisions, but there was no requirement to hold an investigation. Nor was there a set form for such an inquiry until the mid-eighteenth century, when clear guidelines were laid down, though magistrates did follow a more-or-less standard practice, even during the sixteenth and seventeenth centuries.[5]

A precognition is formally the process and (informally, by usage) the outcome of a magistrate interviewing witnesses (and a suspect, if appro-priate), assessing evidence, and summarising an opinion, which informs a decision about whether to take legal proceedings.[6] The imperative with sudden death came more from a need to allay thoughts of foul play, which would require further legal action, rather than to conform to statutory requirements for an inquest. The main aim of the procedure was not (as in England – or France) to identify a 'suspect', but for the conducting magistrate to determine whether the body was a victim of human action; in other words, to discover if a crime had been committed.[7] Even then, the findings could not be founded upon in any subsequent proceedings, as they could with an English coroner's inquest. Any decision to seek a warrant to pursue criminal proceedings over the death was left formally to the sheriff, in practice to any prosecuting magistrate or 'procurator fiscal'.[8] The procurator 'as publick Officer of the Law, represents the Lieges within the Jurisdiction he acts'.[9] A death by accident or suicide required no further proceedings in Scotland unless uncertainty existed over claims to material assets, which would in any case be decided by a civil court.

A Scottish 'inquest' was something quite different from the meaning of the word in connection with an English coroner; it was a way of testing a civil claim by jury. In 1527 the records of the burgh of Stirling contain a reference to the finding of an 'inquist, nane discrepand, that wmquhill [the late] David Wilsoun was nocht the caus of hus awin deid and that he slew nocht him selfe, and that he was wesyit [dying] with siknes, and ane aild waik man, and passand to do his nedis and fell befoir hus awin dour and could nocht recower na help him selfe for waikness, and thar haistely weseit with God'.[10] The marginal key is 'pro Margareta Stevensoun', who must have brought the case as a potential beneficiary of Wilsoun's estate. Sheriff Court records mention such cases only rarely and not as routine inquests in the English sense, but as investigations into disputed cause of death. David Wilson died through illness and misfortune, not suicide. Had it been the latter some of his moveable assets would have been forfeited or 'escheated' to the crown or a lord holding its franchise, for suicide was 'of its nature criminal', even if the case was 'only pursued, *ad civilem effectum* [for civil ends]'.[11] The single entry about suicide in Shetland's 'Sheriff and Justice Court' for the years 1602–4 is: 'Jhone Ollawsoun for hangeing himselff, his guidis and gere [property] escheat'.[12] This was a suicide, with the forfeiture authenticated before a court of record not as a matter of course, but because someone had not accepted it as such.

A dispute over cause of death between living parties lay behind a court case of 1612, which looks like an inquest. The widow of John Elliot in Redden prosecuted the town officers of Hawick (Roxburghshire) on behalf of herself and her 'fyve fatherless bairnes [children]'. She claimed that her husband had died in irons while locked up in a steeple and that the magistrates were responsible. Counsel for the defence argued that the indictment was irrelevant because John or Jock had hanged himself there with his own belt: 'having desperatlie put hands in him self'. The Court of Justiciary asked a jury to ascertain the facts. They heard evidence from two men (following Roman and canon law, proof in Scots law required two witness testimonies), who stated that they were in Hawick marketplace when they heard of the attempted suicide. They went to the steeple, where they found Elliot hanging, still alive; they cut the belt and brought him out, only to see him die shortly afterwards. The jury found the town officers innocent and discharged them, the finding of suicide ending rather than initiating the case.[13]

DOI: 10.1057/9781137381071.0007

The issue of intent and responsibility in the case of suicide did not concern investigators, though the law distinguished between the sane (who might forfeit personal property as felons) and the lunatic (who could not be held to account).[14] In Scots law, it was enough to satisfy certain general principles when conducting an investigation, with judgements about blame or decisions on action made in another forum. Medieval and early-modern English coroners' inquests focused on intention and responsibility when pronouncing on sudden death, and this began to be reported routinely in newspapers from the late-eighteenth century. Scots law had formal structures for assessing mental state, but seldom used them, and the workings of the law of suicide suggest that intention was of secondary importance. In its routine investigative procedures, Scotland never made the transition from the medieval conflation of all types of blameworthy death (including suicide) to a distinction based on state of mind.[15] Instead, responsibility lay with the Court of Session (Scotland's supreme civil court) or another tribunal such as a Sheriff Court that could issue a 'declarator' (statement that a right belonged to the plaintiff). In practice, magistrates formally investigated and recorded only sensational or suspicious cases of suicide that could conceivably have been caused by someone other than the dead person.

Nor did Scottish investigations of death ever decide on deodands, because both the notion and the practice were effectively unknown in Scotland.[16] The King's Council investigated a rare instance early in 1481 when 'The Lords found that the horse whereupon umquhile Thomas Bullock, servitor to James [blank] of Durham, ran in the water of Aven [River Avon, West Lothian] and was drowned, was not escheat to our sovereign Lord; because, by an inquest taken before the Sheriff of Linlithgow, by command of the Lords of Council, it was found that the said Thomas forced the horse with spurs to take the water, and through his own folly and rashness was drowned; and not the horse's fault. And therefore decerned [decided] the horse [was forfeited to] to the said James [blank] of Durham.'[17] Where English coroners had to inquire into any instrument 'inanimate or beast animate' that caused a death, this very idea was alien to Scots law.[18] The Council may have been trying to apply English law to an Englishman, as they sometimes did with other distinctive groups like gypsies.[19]

The Council used the sheriff because he had ultimate responsibility in such matters (and much else). Much later, in his lectures on medical jurisprudence at Edinburgh University c.1830, Professor Robert Christison

DOI: 10.1057/9781137381071.0007

summed up both the historic situation and the contemporary differ-ence between investigations into sudden or suspicious death in parts of Britain: 'a judicial inspection is conducted in Scotland under the warrant of the Sheriff of the county, and in England under that of the coroner'.[20] Indeed, from the twelfth century the power of Scottish sheriffs waxed as that of their English counterparts waned.[21] The Scottish sheriff had the investigative and judicial functions of an examining magistrate and, until the late nineteenth century, responsibility for prosecuting crimes and presenting offenders for trial lay with him.[22] Scotland had 33 sheriffs before 1748 and 27 after that date, the post-1748 officers known more cor-rectly as sheriffs-depute.[23] In practice magistrates or procurators fiscal of particular courts conducted investigations, making Scottish procedures as localised in their own way as English. England had approximately 330 coroners in the early nineteenth century.[24]

A lesser magistrate such as a Justice of Peace (an officer introduced on the English model by legislation of 1587/1609 and appointed by the Scottish Privy Council), bailie or court officer normally did the work.[25] Local officials might also deal with the administrative aftermath of judi-cial investigation. For example, when Thomas Dobbie drowned himself in February 1598, Edinburgh burgh council ordered a bailie and his clerk to 'mak inventar of the guids … and arreist the sam and intromett with [take charge of] his keyis'; the same happened with the goods of John Johnstoun in January 1603.[26] In this and other instances, Edinburgh mag-istrates held the formal hearing to establish suicide in a bailie court rather than a Sheriff Court.[27] Some magistrates admittedly acted as sheriffs and others were ultimately answerable to the sheriff. Edinburgh's magistrates had always had extensive powers, the provost as sheriff and the bailies as deputies.[28] Until the mid-eighteenth century, towns such as Haddington also claimed exemption from the sheriff's jurisdiction. Yet long before then local magisterial investigation was the norm, even in towns without shrieval jurisdiction, and it was not until the mid-eighteenth century that officials formally called procurators fiscal routinely took precognitions. The magistrate or procurator fiscal investigating a death had the closest connection to its circumstances, for the location of the body mattered much less than in England, where coroners had to see the body (and had to be resident within their jurisdictions).[29]

Precognitions were the basic tool for investigating sudden or suspi-cious death in Scotland no later than the sixteenth century. A rare early example concerning suicide comes from the burgh of Edinburgh during the 1640s, when a bailie and town clerk questioned several people about the death of a woman lodger in the house of the widow Beatrix Tailyer, who lived at the head of the Canongate. They recorded the findings in papers annotated: 'Declaratione of severall persons anent the strangling of Katherine Anderson, 19 February 1647'.[30] Anyone who might have known of a dead person's circumstances and state of mind could be called to give evidence. In this case, four people, whose evidence was written down, approached her death from different standpoints. Beatrix said Katherine had previously been in service with Agnes Glen in Leith and owed her the substantial sum of 900 merks (£600 Scots or about £50 sterling). She had repaid 500 merks, but Glen brought a case against her and threatened her with prison for the remaining 400. Katherine lodged in Beatrix's house for two or three weeks and sold some goods to help pay off the debt.

Beatrix had no inkling of what might happen that morning when they were having breakfast. After eating she went to Mr Patrick Henderson to get a testimonial for her daughter, who was in service with a gentlewoman 'and left the said woman [Katherine] in very good cair [state of mind], reading on the new testament, having only the half door shutt' (and so allowing people to see in). She was away less than an hour, but on returning her daughter met her and told her of Katherine's death after a visitor found her hanging by her garter. A soldier cut her down, explaining 'that she was at that time hott and did gaspe once or twice after she was laid in bed'. Beatrix found her house full of strangers on her return, including the soldier and William Sklaitter, deacon of the 'websters' (weavers), because publicising the discovery of a body allayed suspicions that could alight on the finder and because people usually tried to involve someone in a position of authority before dealing with an injured or dead body. Other witnesses said Katherine was unhappy, though the last person to give evidence, her mother, was adamant that she never gave any hint of wishing to kill herself.

For Edinburgh's magistrates, Katherine Anderson was an obvious suicide, even if the annotation on the bundle of papers they compiled was more ambiguous – though forensically precise. Precognition records like this seldom survive except when some other legal process followed the investigation of a death. In this case, the issue was whether Anderson's assets should be forfeited and who should have claim to them. More usually, documents from these investigations survive only if evidence of foul play existed. The following is an example, one of only two cases of suicide listed in the National Archives of Scotland's online index to early-nineteenth-century precognitions for Scotland's highest criminal tribunal, the High Court of Justiciary. A passer-by found the drowned body of Janet Houston in the canal at Paisley on the morning of Sunday 28 March 1816. There were hints of foul play, but evidence of her prior disordered state of mind and her threats to end her life led the procurator fiscal to annotate the packet of documents for her case: 'there is every reason to believe that this woman committed suicide by drowning herself and there is no ground for any judicial procedure'.[31]

State of mind was indirectly relevant in classifying the death because the procurator fiscal had to offer a plausible explanation for the demise, but it had no direct bearing on anything else. In Janet Houston's case, the procurator mentioned it simply to preclude the need for further proceedings. Because the process was private, Janet's death never made it into the Glasgow newspapers. In contemporary England, inquest proceedings were widely reported both by word of mouth and in newspapers. By contrast, reporting of one type of sudden death, suicide, was only about one-third as common in Scottish newspapers, c.1750–1830. An unusual (but characteristically cryptic) report that did make it into the press told of a young man called Stewart Spence, a legal clerk or 'writer', who cut his throat in the open street in February 1772. The report aimed to still rumours of murder: 'From a precognition taken, we can assure the public, he was the unhappy instrument of his own death'.[32]

Magistrates followed more or less regular procedures, shown in the preceding examples, from an early date, though formal instructions about the conduct of precognitions originate in the eighteenth century. Appended to volume two of Peebles Sheriff Court Criminal Sederunt (Minute) Book is the following guidance: 'When any dead body is found with the appearance of violence upon it, or where any person dies, and is suspected to have died by violence, the dead body must be opened, and also the head, and a report made of the cause of the death of the person by physicians and surgeons'.[33] This instruction first came formally from the Crown Office in 1765, issued by the crown agent and depute clerk of Justiciary.[34] Although this document may describe what had been happening for centuries, it had no statutory sanction and no legal authority other than as a recommendation to investigating authorities.

Action *might* result because the limits on precognitions were probably vague until an important case in 1770, when a former customs officer called Mungo Campbell killed himself while awaiting execution for

DOI: 10.1057/9781137381071.0007

murder. Legal counsel for his family got wind that Edinburgh's magistrates intended to have his body given to anatomists, as his sentence had specified under the 'Murder Act' of 1752 (25 Geo. II, c. 37). 'They therefore resolved to attempt to rescue the body from the intended disgrace; and one of them was deputed to represent the matter to the magistrates. He found one of them, and their clerks, in the council-chamber, taking a precognition; which they told him they always did in such cases.'[35] Counsel condemned the procedure as 'inhumane and unjust; that no man (dead or alive) ought to be found guilty of a crime, without fair trial, which this precognition was not; and that the counsel for the prisoner could find no law authorising such procedure.'[36] This opinion prevailed and helped to identify the judicial limits of precognitions.

The foregoing analysis indicates a tendency in Scotland to employ certain procedures. Precognitions *may* have been routinely conducted in cases of suicide (even if the Lord Advocate, the senior crown officer in Scotland, did not take over paying procurators for conducting them until 1776), but, because no process followed, the papers seldom survive.[37] There nevertheless remained a crucial difference between English and Scottish investigative procedure. An English coroner had to hold an inquest if anyone asked for it and a verdict of wrongful death was, of itself, sufficient authority for bringing criminal proceedings. Even in the early nineteenth century, by contrast, there was no obligation to investigate any death in Scotland and an enquiry was at the discretion of magistrates. An early-nineteenth-century summary of procedures north and south of the Border stated: 'The power of enquiring into the causes of Sudden Death, which is the special duty of the English Coroner is seldom exerted by the sheriffs of Scotland, and no inquest is usually held upon the discovery of a Dead body, unless when suspicion of Murder exists.'[38]

Unlike England, there was no obligation on magistrates to return findings to a central court of record until the nineteenth century, when reports from procurators fiscal began routinely to be collected and collated by the Crown Office.[39] At this date magistrates enquired into any death not obviously natural, taking the testimony of witnesses and/or acquaintances, a statement from the apparent perpetrator if there was a hint of foul play, and advice from one or more medical practitioners about the cause of death.[40] The resulting report was usually very short and free of technicalities.[41] After hearing the report, magistrates held an oral discussion in which reasons for the opinions might be asked,

DOI: 10.1057/9781137381071.0007

thereafter transmitting notes of the precognition to the Crown Office in Edinburgh to help crown counsel determine if there was a case to be answered.[42]

Scottish investigative procedures were at least as rigorous as English, but not as consistently invoked. They resulted from a more discreet judicial system in which a public official pursued the truth by informing himself in closed session. Even when investigators sought medical testimony, there was no obligation to establish cause of death provided criminal proceedings were ruled out. The circumstances of the case were the principal determinant of action. Financial considerations also limited willingness to hold formal inquiries, because the cost of inquests was subject to assessment by landowners until the early Victorian era. They frowned on any hint of unnecessary expense and 'policed' sheriffs through the threat of not reimbursing an investigation's costs. This was still an issue when George Salmond, a Glasgow procurator fiscal, wrote to Thomas Tancred in June 1841.

> Though I am not aware of any instance occurring since my appointment as fiscal in 1816, where investigation was omitted, yet I have no doubt that from the indifference of persons particularly in the country districts, to report such cases unless very flagrant and there being no compulsitor [legal requirement]; instances may have occurred where no report has been made to me & of course no investigation made. On this account your Establishment of a Coroner's Inquest is invaluable in rendering investigation *imperative & instant* and as giving encouragement to every one to inform as soon as possible. But in Scotland, the chief obstacle would be, as it always has been, to get money to remunerate the informer & witnesses, the jurors etc. & especially as to cases where the cause of death has been *accidental & self-evident by inspection*. ... every thing here of that kind is grudged & if any expence were charged as to a case where the cause of death was plain, such as from a fall off a house or the like, it would be refused.[43]

Cost (and perhaps common sense) looms large in this account. The system of public prosecution in Scotland had to balance the need to investigate suspicious deaths with accountability for spending money from the public purse.[44] Concern about allegations of profligacy may have encouraged procurators to find a death accidental: a Scottish autopsy required two doctors, both requiring fees. Procurators fiscal had to fund a medico–legal inquiry out of a limited supply of 'rogue money' (an assessment towards crime prevention). Only if the case went up to the Justiciary Court would their costs be met from central funds.

DOI: 10.1057/9781137381071.0007

Financial considerations may also explain why the rate of post mortems in Scotland was lower than England. In nineteenth-century England and Wales, coroners' inquests considered roughly one death in twenty compared with just over one-third of all deaths at the present time.[45] Between 1848 and 1857, the Crown Office investigated approximately 500 'reports of deaths' annually in Scotland out of roughly 20,000 deaths each year, giving a rate half that in England.[46] Some writers claimed that authorities generally took cause of death for granted.[47] This approach may be one reason, until late in the twentieth century, rates of recorded suicide in Scotland were much lower than those in England and Wales.[48] Of course, English county authorities were not free spenders. Perhaps influenced by examples of Scottish economy, magistrates told constabularies in Northumberland, Durham and the West Riding in the 1850s to save money by not reporting every sudden death to coroners.[49]

The last century or so has seen a refinement of procedures, which had been followed in earlier times, and their clarification by legislation. Fatal Accident Inquiries have been held by Sheriff Courts since 1895, when an act introduced mandatory public inquiries into work-related deaths, before a sheriff and jury (an act of 1906 added jurisdiction over sudden or suspicious deaths). In modern Scotland 42 procurators fiscal perform the functions of coroners under the Fatal Accidents and Sudden Deaths Inquiry (Scotland) Act (1976). Police, medical professionals or registrars (introduced into Scotland in 1855) supply them with information, though in theory anyone may notify a death.[50] Police use their initiative in the early stages of an investigation, but they have no authority to act independently of the procurator fiscal and he alone decides on the relevance of information. The procurator fiscal is an investigator, leader of evidence and commentator thereon in the public interest, appointed by and answerable to the Lord Advocate. In the case of sudden deaths now, he or she conducts direct inquiries in person and in private using medical or any other appropriate expertise, rather than holding a public inquest. The Crown Office supervises procedures and makes decisions.

Advocates-depute not only prosecute the serious cases which make up the business of the High Court of Justiciary, but also decide, on the basis of papers sent up by the local procurators fiscal to the Crown Office, whether to institute a prosecution and, if so, in which court. Further, they consider reports from the procurators fiscal about deaths. Deaths could be recorded in the normal way by the registrar, or be designated a cause such as suicide (a classification rather than a verdict), or the

DOI: 10.1057/9781137381071.0007

advocate-depute could instruct a Fatal Accident Inquiry to be carried out before the sheriff (no longer now with a jury) with the procurator fiscal leading the evidence; only in the last was a formal 'verdict' offered. The advocate-depute can then instruct an entry to be placed in the registers stating that death was due to a specified cause, 'per decision of fatal accident inquiry'. As much as anything, this reflects Victorian and later concerns with accurately tabulating causes of death and is a part of 'police', which in Scotland meant a wide-ranging concern with environment, health and order ('polis' or civic government) rather than what the English described as 'police' (constabulary).[51]

Notes

1 Anon., 'Official inquiry in cases of sudden death', *Journal of Jurisprudence* 2 (1858), 273–4. The statement on p. 274 that 'the title of *Coroner* belonged to the Sheriffs by our ancient style' is misleading in implying that the connection was automatic.

2 R. A. Houston, *Punishing the dead? Suicide, lordship, and community in Britain, 1500–1830* (Oxford: Oxford UP, 2010).

3 *The tryall of Philip Standsfield, son to Sir James Standsfield, of New-milns, for the murther of his father, and other crimes libell'd against him, Feb. 7. 1688* (Edinburgh: heir of Andrew Anderson, 1688), 12.

4 Those accused of serious criminal offences also routinely had counsel from the sixteenth century, much earlier than was the norm in England. M. Wasser, 'Defence counsel in early modern Scotland: a study based on the High Court of Justiciary', *Journal of Legal History* 26 (2005), 183–201. J. B. Post, 'The admissibility of defence counsel in English criminal procedure', in A. Kiralfy, M. Slater and R. Virgoe (eds), *Custom, courts and counsel* (London: Frank Cass, 1985), 23–32.

5 R. Clark, *A view of the office of sheriff, in Scotland* (Edinburgh: Bell & Bradfute, 1824), 14–17. Alison, *Criminal justice*, 56–9.

6 Precognitions potentially left a paper trail, whereas the informal investigation, which probably preceded them, usually did not. F. Russell, 'On the procedure in criminal prosecutions in Scotland preliminary to trial', *Journal of Jurisprudence* 14 (1870), 262.

7 A. Joblin, 'Le suicide à l'époque moderne. Un exemple dans la France du nord-ouest: à Boulogne-sur-Mer', *Revue Historique* 129, 1 (1994), 89. Alison, *Criminal justice*, 56–9.

8 Russell, 'Procedure in criminal prosecutions', 264–5. The procurator fiscal may originally have been a financial agent of the sheriff (as the name

DOI: 10.1057/9781137381071.0007

implies), though his curial role seems to post-date the Reformation and to have begun in burgh courts before sheriff courts. W. Reid, 'The origins of the office of the procurator fiscal in Scotland', *Juridical Review* 10 (1965), 154–60. The word *fiscal* (Spanish) or *fiscale* (Italian), meaning a public prosecutor, is found in late medieval Europe, suggesting a Roman law derivation. T. V. and E. S. Cohen, *Words and deeds in Renaissance Rome: trials before papal magistrates* (Toronto: University of Toronto Press, 1993), 17. L. A. Homza (ed.), *The Spanish inquisition, 1478–1614: an anthology of sources* (Cambridge: Hacket Pub Co., 2006), xxii.

   9 From a case of 1741 quoted in J. Finlay, 'Pettyfoggers, regulation and local courts in early modern Scotland', *Scottish Historical Review* 87 (2008), 59.

10 R. Renwick (ed.), *Extracts from the records of the royal burgh of Stirling, A.D. 1519–1666* (Glasgow: Scottish Burgh Records Society, 1887), 31. This meaning of inquest was also an earlier use in English law. I owe this point to my colleague John Hudson.

11 J. Erskine, *An institute of the law of Scotland in four books, in the order of Sir George Mackenzie's institutions of that law* 2 vols (Edinburgh: John Bell, 1773), IV.IV.46.

12 G. Donaldson (ed.), *The court book of Shetland, 1602–1604* (Edinburgh: Scottish Record Society, 1954), 23. There is a single similar case of a declaratory action (asserting right) among the published records of Aberdeenshire Sheriff Court, 1598–1649. D. Littlejohn (ed.), *Records of the Sheriff Court of Aberdeenshire* 3 vols (Aberdeen: New Spalding Club, 1904–7), vol. 2, 351.

13 R. Pitcairn, *Criminal trials in Scotland, from A.D. MCCCLXXXVIII to A.D. MDCXXIV* 3 vols (Edinburgh: William Tait, 1833), vol. 3, 219.

14 Forfeiture for felony, which included suicide, lapsed during the eighteenth century, but was not abolished in Scotland until 1949. *An introduction to Scottish legal history*, 445.

15 S. K. Morrissey, *Suicide and the body politic in imperial Russia* (Cambridge: Cambridge UP, 2007), 27–9.

16 Lord Cooper, 'Curiosities of medieval Scots law', *Proceedings of the Royal Philosophical Society of Glasgow* 69 (1944–5), 54. A. D. M. Forte, 'The horse that kills: some thoughts on deodands, escheats and crime in fifteenth century Scots law', *Tijdschrift voor Rechtsgeschiedenis* 58 (1990), 95–110. B. S. Jackson, 'Liability for animals in Scottish legal literature: from the *auld lawes* to the sixteenth century', *Irish Jurist* 10 (1975), 334–51. Adam Smith's discussion of 'deodat' is about England. A. Smith, *Lectures on jurisprudence* edited by R. L. Meek, D. D. Raphael and P. G. Stein (Oxford: Oxford UP, 1978), 116–17.

17 W. M. Morison, *The decisions of the Court of Session … in the form of a dictionary* 42 vols (Edinburgh: Bell & Bradfute, 1801–7), supplement 1, 113.

This judgement follows the example in *Quoniam Attachiamenta*. T. D. Fergus (ed.), *Quoniam Attachiamenta* (Edinburgh: Stair Society, 1996), ch. 35 [p. 203]. Walker, *Legal history of Scotland*, vol. 3, 729–30.

18   Wellington, *King's coroner*, 14.

19   Goodare, *Government of Scotland*, 263–5.

20   Edinburgh University Library Special Collections, Dk.4.57, 'On medico–legal inspections', f. 2v. R. W. Renton, 'The investigation of cases of sudden death in Scotland', *Juridical Review* 5 (1893), 167–74. The official involved is variously described as the sheriff, procurator fiscal or solicitor-general.

21   Morris, *Medieval English sheriff*, 238–9. M. J. Braddick, *State formation in early modern England*, c.1550–1700 (Cambridge: Cambridge UP, 2000), 30–1.

22   G. Little, 'Local administration in Scotland: the role of the sheriff', in W. Finnie, C. M. G. Himsworth and N. Walker (eds), *Edinburgh essays in public law* (Edinburgh: Edinburgh UP, 1991), 295–314. Walker, *Legal history of Scotland* vol. 3, 554.

23   J. Wallace, *The sheriffdom of Clackmannan ...* (Edinburgh: James Thin, 1890), 20, referring to the reign of George II. A. E. Whetstone, 'The reform of the Scottish sheriffdoms in the eighteenth and early nineteenth centuries', *Albion* 9, 1 (1977), 61–71.

24   Anderson, *Suicide*, 15. There are currently 259 coroners' districts in England and Wales.

25   R. S. Tompson, 'The justices of the peace and the United Kingdom in the age of reform', *Journal of Legal History* 7 (1986), 273–92. S. J. Davies, 'The courts and the Scottish legal system 1600–1747: the case of Stirlingshire', in V. A. C. Gatrell, B. Lenman and G. Parker (eds), *Crime and the law: the social history of crime in western Europe since 1500* (London: Europa, 1980), 132–4. There is a thesis, submitted to the University of Glasgow in 1934, but refused, that deals in a curious way with this topic. W. A. McNeill, 'The administration of Scotland: with special reference to the Justice of the Peace system 1603 to 1625'. The only extant copy is in Edinburgh University Library Special Collections, SD 7934.

26   Edinburgh City Archives, SL1/1/10, f. 177v (8 March 1598). SL1/1/11, ff. 111v–112.

27   Edinburgh City Archives, Moses bundles 11, no. 452, item 23 (1647).

28   *Charters and other records relating to the city of Edinburgh, A.D. 1143–1540* (Edinburgh: Scottish Burgh Records Society, 1871), 157–65. Pitcairn, *Criminal trials*, vol. 1, 129–30.

29   I. H. B. Carmichael, *Sudden deaths and fatal accident inquiries: Scots law and practice* (1986. 2nd edition, W. Green: Edinburgh, 1993), 7.66.

30   Edinburgh City Archives, Moses bundles 11, no. 452, item 23.

31   NAS AD14/16/13. AD14/21/208.

32   *Glasgow Journal* no. 1587 (18 February 1772). For the original precognition see Edinburgh City Archives, McLeod Bundles, DO113, item 30 (1772). Anon.,

'Coroners' inquests for Scotland (Fatal Accidents Inquiry Bill)', *Scottish Law Review* 9 (1893), 228.

33 NAS SC42/23/2, item 3. The volume covers 1758–92, but the regulations are similar to those covering other aspects of the Sheriff Courts published in 1749. *Regulations proposed to be observed in the Sheriff and Stewart courts of North Britain* (Aberdeen: J. Chalmers, 1749).

34 'Rules to be Observed in taking Precognitions'. This printed broadsheet is bound with a collection of similar papers and pamphlets relating to Aberdeenshire. It can be found reprinted in Clark, *Office of sheriff*, 23–7. D. Hume, *Commentaries on the law of Scotland, respecting crimes* 2 vols (1797. 3rd edition. Edinburgh: Bell & Bradfute, 1844), vol. 2, appendix X, 535. NLS X.225a.1(2). The Crown Office was and is a government department created to administer the criminal law, comprising the Lord advocate, the solicitor general for Scotland, the crown agent, and several advocates-depute (also called Crown Counsel). A new set of rules appeared in 1824 and further revisions have since been made.

35 *Arguments and decisions in remarkable cases before the High Court of Justiciary and other supreme courts in Scotland. Collected by Mr MacLaurin* (Edinburgh: J. Bell, 1774), 530.

36 *Idem.*

37 As in the case of William Pollock. *Further particulars about William Pollock, who hanged himself in the jail of Edinburgh, on Monday the 20th March, 1826, with his dying declaration in a letter to a gentleman the night before his death* (Edinburgh: William Robertson, 1826). A. E. Whetstone, *Scottish county government in the eighteenth and nineteenth centuries* (Edinburgh: John Donald, 1981), 20.

38 Northumberland Record Office ZRW/468, 'Respecting the office of coroner in Scotland & Northumberland' (1812).

39 NAS AD12/11–12. AD13/1–459.

40 For rare examples see Edinburgh City Archives, Moses bundles 11, no. 452, item 23 (1647) and McLeod Bundles, DO113, item 30 (1772). For a murder investigation see R. A. Houston, 'New light on Anson's voyage, 1740–1744: a mad sailor on land and sea', *Mariner's Mirror* 88, 3 (2002), 260–70.

41 NAS AD13/1–459 (1825–55).

42 Edinburgh University Library Special Collections, Dk.4.57, 'Medical reports, medical evidence', ff. 3–3v.

43 Northumberland Record Office NRO 530/2/6.

44 Alison, *Criminal justice*, 17–24, rehearses the countervailing forces in decisions to prosecute. M. A. Crowther, 'Scotland: a country with no criminal record', *Scottish Economic and Social History* 12 (1992), 84. M. A. Crowther, 'Crime, prosecution and mercy: English influence and Scottish practice in the early nineteenth century', in S. J. Connolly (ed.), *Kingdoms*

DOI: 10.1057/9781137381071.0007

united? Great Britain and Ireland since 1500: integration and diversity (Dublin: Four Courts, 1999), 225–38.

45 I. A. Burney, *Bodies of evidence: medicine and the politics of the English inquest, 1830–1926* (Baltimore: Johns Hopkins UP, 2000), 3.

46 NAS AD12/11–12. The difference remains to this day. In England and Wales, roughly 25 per cent of deaths are referred to a coroner whereas in Scotland only 13 per cent are sent to the procurator fiscal. At 9 per cent, the Scottish autopsy rate is a third of that in England and Wales.

47 J. Craig, *On the office of the coroner; and on medical evidence in the preliminary investigation of criminal cases in Scotland* (Edinburgh: Sutherland and Knox, 1855), 22–4.

48 E. Stengel, *Suicide and attempted suicide* (Harmondsworth: Penguin, 1964), 19. Renton, 'Sudden death', 167. Crowther, 'Criminal record', 84. O. Ross, 'An investigation of official suicide statistics in Scotland compared with those in England and Wales', (University of Edinburgh M.Phil., 1974).

49 Anderson, *Suicide*, 20n. Sim and Ward, 'Magistrate of the poor', 253-7. M. Jackson, 'Suspicious infant deaths: the statute of 1624 and medical evidence at coroners' inquests', in M. Clark and C. Crawford (eds), *Legal medicine in history* (Cambridge: Cambridge UP, 1994), 64, suggests that, in the absence of medical input before 1836 (6 & 7 Wm IV c. 89), English coroners only investigated deaths with 'obvious marks of violence', yet the rate of investigation was still higher than in Scotland. M. B. Emmerichs, 'Getting away with murder?', *Social Science History* 25 (2001), 93-100.

50 A. V. Sheehan, *Criminal procedure in Scotland and France* (Edinburgh: HMSO, 1975), 220–1. Carmichael, *Sudden deaths*, 7.66. *Death and the procurator fiscal* (Edinburgh: Crown Office, 1998).

51 B. White, 'Training medical policemen: forensic medicine and public health in nineteenth-century Scotland', in M. Clark and C. Crawford (eds), *Legal medicine in history* (Cambridge: Cambridge UP, 1994), 145–63. D. G. Barrie, *Police in the age of improvement: police development and the civic tradition in Scotland, 1775–1865* (Cullompton: Willan, 2008); 'Anglicization and autonomy: Scottish policing, governance and the state, 1833 to 1885', *Law and History Review* 30 (2012), 449–94.

DOI: 10.1057/9781137381071.0007

3

Scottish Coroners: Origins and Development of the Office to c.1500

Abstract: *Scotland had coroners, but their roles were quite different from their English namesakes. This chapter finds their origins in a mix of Celtic or Gaelic, English, British and Scottish institutions of government, justice and peace-keeping, which gelled into a coherent system for the administration of justice during the fourteenth and fifteenth centuries. Where English coroners were independent judicial office holders, Scottish coroners were judicial agents or executive law court officers: men of action who arrested suspects and seized goods on behalf of the king's judges. They dealt with the living rather than with the wrongfully dead. They also had quasi-military functions and some powers of summary justice when maintaining law and order. The chapter deals with the location of coroners within Scotland, their social status, remuneration and functions, making extensive use of historical documentation.*

Houston, R. A. *The Coroners of Northern Britain c. 1300–1700*. Basingstoke: Palgrave Macmillan, 2014. DOI: 10.1057/9781137381071.0008.

If modern historians are vague about the Scottish coroner, contemporaries recognised his importance. The office certainly existed in the Middle Ages and features in many different types of documents from charters to poetry.[1] Some of the great Victorian authorities familiar with these sources ventured educated guesses about the coroner's origins, functions and fate. One was advocate and antiquary Cosmo Innes. 'The name was derived ... from this officer having cognizance in the pleas of the Crown – *placita coronae*. At one time, the functions of crowner were very high, both in England and Scotland, and seem to have been co-extensive with the sheriffdom. I do not know at what period the coroner's duty in England was restricted to what it is at present. The office went early out of use in Scotland.'[2]

Contemporary English medico–legal writers added equally brief, if slightly better grounded, comments as contributions to debates about how to improve the investigation of sudden or suspicious death. 'Coroners or "Crowners" are mentioned in many old Scotch Statutes, but the office was either abolished or fell into disuse in Scotland, probably in consequence of the Succession War and the French connection, and now the duties of a coroner as of a public prosecutor in criminal cases ... are performed by a Crown official, styled the procurator fiscal, who is usually a Scottish legal practitioner of some standing ... appointed by the sheriff of the county, and in cities or towns by the magistrates.'[3] Correct about the investigation of sudden death by procurators fiscal (by the time of George III lawyers in the public service appointed by local judges), but wrong in implying that coroners once did this in Scotland, the writer glossed over the nature and functions of the office. Usually mentioned in a few lines, if at all, the Scottish coroner remains to this day an 'office of some obscurity' to legal and other historians alike.[4]

It is conventional to assume copying from England, a transplant similar to that used in the legal and judicial assimilation of Wales in the 1280s.[5] After the Welsh conquest, Edward I also instituted profound changes to the English law of homicide to make it easier for eligible offenders to obtain routine pardons and more difficult for others to avoid royal justice. As part of this design, all stages of examinations into suspicious deaths involved coroners. Thereafter they were crucial agents in the collection of fines, amercements and forfeitures.

Edward wanted to take direct control of Scotland's system of justice when he occupied parts of the country after 1296. Thus, the first use in Scotland of the word coroner with named individuals comes from

Edward's time. Sir Robert Boyd of Noddsdale worked for the English as coroner of Ayrshire and Lanarkshire in 1304, whereas William Lekprevik was 'coronator' of the lordship of Kilbride soon after.[6] English influence around this time may create an impression of transmission between the countries. Robert I's charter to Berwick-upon-Tweed (1320) gave the appointment of coroner for the burgh to the justiciar of Lothian; this seems to follow Edwardian charters and also that of Henry III to London in 1268.[7] It is not the only example of eliding prescriptions. Legal writer Sir James Balfour quoted the *Leges Forestarum* to support the assertion that coroners or other crown officers should oversee the safe-keeping of 'sea-wrak', though this too looks very like the English 'escheit of the sea' mentioned in 3 Ed. I, c. 4.[8]

Judicial texts seem to confirm the English origins. The entry 'coroner', in Jacobean jurist Sir John Skene's *De Verborum Significatione*, says that 'Crouner inquires be ane inquest anent murther and slaughter done, and committed quietlie.'[9] This seems to be less a description of Scottish practice than a borrowing from English usage (as, of course, is the *Regiam Majestatem*, which his text explicated[10]), for the authorities cited are English and the procedure likewise: '[t]he quhilk inquisition suld be taken in the hie streites, or in open places, in *corona populi* [before the people assembled]'. The influence of English procedures is also seen in the law of rape, which borrows from the *Regiam* in requiring abused women to 'pass fordwart on the king's way to the schiref [sheriff] of that schirefdome, or to the coroner, gif he may be had' and to present evidence to him.[11] When documented, from the sixteenth century, formal investigation of sudden death in Scotland actually took place in private; matters only became public with an accusation against a living person or in case of a dispute over cause of death. Other than public requests to identify a body or to trawl for information, the procedures outlined by Skene do not seem to have been followed, suggesting that he presented what he thought the officer could be doing, not what he did.[12] In all likelihood, there was already an infrastructure for dealing with sudden death (and most other judicial matters) in Scotland long before Edward's time, leaving few meaningful opportunities for officials like coroners to do a job long carried out by other officials like mairs and serjeants (discussed later).

The same is true of England, which had serjeants of the peace. They had extensive powers of summary justice, but their main duties focused on justice eyres. In the borough of Chester the palatine Earl had a master

DOI: 10.1057/9781137381071.0008

serjeant of the peace. Here in Cheshire and in Shropshire and Lancashire too, the hundreds, wards or wapentakes (a Norse word used in parts of England once settled by the Danes) also had their own serjeants and in Cheshire the Earl's barons could appoint serjeants, as they could in the baronies of Northumberland.[13] Thus, there were county serjeants as well as baronial and franchisal ones. Vivian de Davenport had, for example, a grant of the hereditary office of master serjeant of the peace of Macclesfield from Ranulf, Earl of Chester, c.1217–26.[14] Serjeants of the peace subsisted on fees, perquisites and free quarter from the communities they policed (later commuted to cash payments) and by seizing certain assets belonging to the felons they caught and presented.[15]

Serjeants were important because there was no view of frankpledge north of the Humber or on the Welsh marches, frankpledge being a system where a group of men in a township had the responsibility for capturing offenders and presenting them to the authorities. There were, however, other aspects of a national peace-keeping system, such as the hue and cry, in Cheshire and Northumberland. In Ireland too early coroners do not seem to have recorded the tithing (a subdivision of a hundred) responsible for fugitive felons.[16] Serjeants mattered more in areas where tithings were not held responsible for law-breaking; the importance in the same regions of Britain (and Ireland) of franchisal jurisdictions and of holding lords accountable for the actions of their subordinates through sureties may also indicate different priorities in peace-keeping.[17] This may, for example, explain why Prescot (Lancashire), whose charter of 1447 allowed for the appointment of a coroner, has no record of one until 1575.[18]

A slightly later example from the north suggests similar priorities in peace-keeping. In 1615 Elizabeth Kersopp, a customary tenant on the manor of Wark in Northumberland, committed 'murder upon herself by strangling herself'. Lord Howard de Walden claimed her lands were forfeit to him as lord of the manor and keeper of Tynedale. Elizabeth's heirs refused to allow the Lord's new tenant to occupy the lands and he prosecuted them before the Council of the North. The Council seems to have found for the Kersopps, who claimed that because of their Border tenures they were not subject to forfeiture for felony. Howard de Walden then pursued the Kersopps before Star Chamber and argued his case before the Privy Council in 1616–19. Where in earlier times the Council of the North supported tenant right by Border service, the councillors in London upheld the idea that lords should be accountable for their tenants in the 'Middle Shires', which meant reducing the latter's privileges.[19] Of course, government policy at this date was to turn the Borders into the 'Middle Shires', but the case highlights the distinctive social and administrative makeup of the north.

The first regular mentions of Scottish coroners in the Register of the Great Seal and elsewhere come from the time of David II (1329–71), paid for by owners anxious to safeguard their privileges.[20] There is, on the other hand, evidence of the coronership earlier in Scotland, during William I's reign (1164–1214), documented as part of an attempt to regularise the Celtic, Anglo-Saxon and Norse offices of Toiseach, Thegn and Hersir. The assize of William I (c. 30) states that attachments (seizures) at provincial courts can be made 'per seriandum [serjeants] vel [or] coronatorem vel tosordereh', but this seems illustrative or permissive rather than descriptive.[21] More concrete evidence of the office's existence before Edward's time comes from two sources: a description of (or perhaps a set of recommendations for?) the Scottish king's household and government, dated by historian Geoffrey Barrow to c.1292, saying that justiciars should appoint and be responsible for coroners, and the provision in Edward I's ordinances of 1305 for the three main officers of state to appoint coroners if the present incumbents were found unfit 'unless the latter hold by charter', in which case the king's chancellor had to be consulted.[22] The best that can be said is that the word 'coroner' is Anglo-Norman. The office gradually took the place of king's or sheriff's serjeants as keepers of the pleas of the crown in hundreds, wards or wapentakes across most of England from 1194 and it acquired new, more extensive and independent functions as well.[23]

The similarities between late medieval Scottish coroners, and a variety of officers known by the same name – as well as others that had different appellations – reminds one of historian James Campbell's (confessedly controversial) statement that, when looking at elements of the organisation of *Scotia* south of the Forth, 'it is almost as if there were two Englands and one of them is called Scotland'.[24] His inspiration here was Maitland, who wrote to sheriff George Neilson in October 1898: 'Scotch [sic] medieval law is to me so French, so Norman – and the change from English to Scottish is not sudden at the border, but is "mediated" by the condition of our four northern counties, which seem to me the Frenchest part of England. ... It seems as if the later infusion of French jurisprudence met a kindred element in Scotland that had been there a very long time.'[25] Maitland's primary interest lay in land law, though he

DOI: 10.1057/9781137381071.0008

also observed that kingship was weaker, lordship stronger, in northern than southern England, and that extensive privileged areas existed in the north, parallel to French principalities, that were absent from southern England.

Yet if one school of thought looks to England (or Normandy) for the origins of coroners in Scotland, replicating during the twelfth century the Anglo-Norman administrative model of sheriffs, justiciars and so on, another will just as predictably search for its indigenous roots, following the line that new terminology obscures continuity: 'old functions were made to serve in new surroundings' and with different names.[26] It should, indeed, already be clear that the study of Scottish coroners cannot be seen simply as a wholesale transplantation of the English system, judged as a success or failure only by reference to the trajectory of the office south of the Border. Nor was it just a continuation of serjeants or keepers of the peace, as found in England, Wales and south-west Scotland in the twelfth and thirteenth centuries. Nor, for that matter, was there a common 'Scottish' or indeed 'Celtic' 'system', however much the broad functions may have been the same. Instead, there was a variety of both local terminology and specific functions among late-medieval offices. An analogy is the bewildering variety of local weights and measures in Scotland well into the early modern period.[27]

There may indeed have been an older equivalent of the coroners, serjeants, and keepers north of the Forth: the *toíseachdeor* or *toschachdor*. Indeed the Scottish coroner more closely follows his functions (and those of both early English coroners and their predecessors) than he does the later English figure.[28] The toíseachdeor is a mysterious officer, mentioned from time to time from the twelfth to the sixteenth centuries. The term 'toíseach' means 'leader' in a non-specific sense. In Ireland *taisech* or *taoiseach* is used for the man in charge of a former petty kingdom when it lost its sovereignty to a provincial king: 'a vassal-chieftain'.[29] Meanwhile in Gaelic notes about land grants contained in the Gospel Book of the monastery of Deer (now Old Deer, Aberdeenshire, c.1150) it is used for the heads of kindreds.[30] The second element, 'deor', seems to be the word usually translated as 'base' or sometimes 'unfree', but also as 'stranger' (from which the idea of a fugitive derives). 'Deor' most frequently appears in contrast to 'saor' ('free') in legal discussions of clientage: a 'free client' is a freeholder who enters into temporary relations with a lord in return for a grant of cattle or equivalent; a 'base' client is entirely dependent upon his lord economically.[31] Thus a 'toíseach deor'

DOI: 10.1057/9781137381071.0008

is someone in authority who owes his position to delegation from above (probably from a king or *mormaer*) rather than affirmation from below (as would be the case with a Clann chief).[32]

In the past some scholars have hypothesised that 'toíseach' is the Celtic equivalent of 'thane', equating thanes, mairs and *toíseachs* as stewards in charge of running estates held of the crown and acting as the king's local agents (as opposed to *mormaír* or earls, who were provincial military leaders and enforcers of royal justice).[33] This is plausible, but not demonstrable and the *toíseachdeor* was probably an officer of the crown or a great Earl with similar, but not identical roles to the mair or serjeant, charged not only with administration, but also with consolidating lordship in a defined area – possibly a multiple estate.[34] This included military force if necessary and serjeants had the power of summary justice; that they and the coroners over them had licence legally to 'sorn' (seek quarter) and to 'ransel' (search for stolen goods) suggests an active, peripatetic role.[35] Historian Gilbert Márkus has ventured that the word *toíseachdeor* derives from *toísech daortha* meaning 'chief of capturing' or even 'tracker of thieves'.[36]

Informed of a man suspected of stealing a cow, Sir Duncan Campbell of Glenorchy (Argyllshire) told his son to seek him out 'with all diligence... For I wold nocht that my lordis coronar is start befoir me in sic caiss'.[37] This may indicate independent action by coroners and indeed some level of initiative must have been essential in a system of justice where identifying and apprehending suspects was seldom easy. Given the usual assumption that agents of the centre would be met by opposition from landed interests in localities, this remark also suggests that overlaps between franchisal coroners kept lairds (lesser lords) on their toes in peace-keeping; in this case the Earl of Argyll is probably the lord referred to. Coroners' main role was protecting the king's financial interests so he could profit from potentially lucrative revenues like the confiscation of the property of felons and the forfeiture of sureties, but their job was also to bring the king's justice into the localities. Unpopular they may sometimes have been, but they could also be appreciated as they worked with other agencies of local law and government to uphold the king's peace. They may have been especially important in cases where kin could not operate or were compromised, by determining if a murder had happened and identifying suspects.[38]

Being a *toíseachdeor* made a statement about affinity and being a serjeant too may have involved jurisdiction over a kindred, whereas being a coroner signified appointment to a function within the king's legal system, dealing with reserved royal justice.[39] Thus, there was something new about the coroner, who summoned before the justiciar *ex precepto Regis* (on the orders of the king), whereas most other courts with delegated, limited jurisdiction expected suitors there by default. He was an agent of expanding royal justice, as evidenced in the elaboration of the 'pleas of the crown' and the development of the process of presenting defenders during the fourteenth and fifteenth centuries. Maitland wrote that 'New needs are being ever and anon met by new devices.'[40] If the offices of *toíseachdeor* and coroner had been the same in some earldoms, they were different in the royal sheriffdoms of the fourteenth century and beyond. The *toíseachdeor*'s role was nevertheless indirectly a public one too, for (under the *toíseach*) he may have recruited troops for the crown's 'common army' rather than merely enlisting from a group of kin on behalf of a local lord, himself obliged to help the king defend the realm.[41]

Late-sixteenth-century writers equated or conflated the offices because they thought (or wished to promote the idea) that Scotland had had a uniform political and judicial past. The judge David Chalmers (c.1533–92) thought ayres very ancient: '*Et pour y amener les malfaicteurs à obeir à la iustice, les ministres dicts en la langue Irlandoise Touschediracht, en Escossois, Cronars, & en François sergens, furent creez*' (the three named officers each presented wrongdoers before justice). The marginal key is '*Chercheurs ou cronars*' (searchers or coroners), an interesting parallel with the Welsh *ceisiadau* (triers, seekers or, more pejoratively, satellites, catchpoles, inquisitors or extortionists).[42] Skene thought the 'tocheoderache' 'ane office or jurisdiction, not unlike to a bailliarie, especially in the Iles and Hielandes'.[43] Skene tried to distinguish coroner from *toíseachdeor*, but actually confounded them.[44] In his 1609 edition of the *Regiam Majestatem* he also referred to a fourteenth-century grant of the '*officium serjandie comitatus de Carrik, quod officium Toschadorech dicitur, vulgo* ane mair of fee' (office of serjeant in the county of Carrick, also known as *toíseachdeor*, or more commonly mair of fee).[45] The early mair's function leant towards tax collection (including 'cains' or food renders to the crown) and thus, in a debased form, it was still used to describe an estate steward or bailiff in the nineteenth-century Highlands and Islands.[46] Mairs seem to have had more restricted geographical remits and functions than most coroners or *toíseachdeors*, with correspondingly lower status, acting as executors of summonses and 'diligences' (the equivalent of an English 'writ of execution' against a debtor). There were, for example, mairs for

DOI: 10.1057/9781137381071.0008

several districts of Mar, but only one '*toschederach*' of Mar in the 1450s.[47] As implied by the Carrick example, mairs of fee may have been closer socially to coroners or *toíseachdeors*. Seventeenth-century charters to the Earl and Marquis of Argyll equated the 'sergeandry' of north and south Knapdale with mair of fee.[48]

The legacy of the military function of *toíseachdeors* in consolidating territories may perhaps be found in the role expected of coroners in the fifteenth century. In his Scotichronicon, Walter Bower describes a piece of exemplary justice supposedly meted out in Wester Ross soon after the death of Robert I. Thomas Randolph, as justiciar in the Highlands, 'sent his official coroner on ahead to Eilean Donan with an armed force to arrest lawbreakers in accordance with enrolled indictments. This official pursued fifty of them; and because they resisted arrest, they were slaughtered by their pursuers; and the walls were adorned with their heads fixed to poles and sticks before the judge's arrival at Eilean Donan'.[49] Writing in the 1440s, Bower, an abbot as well as a historian, perhaps displayed his preoccupation with law and order at a time when some thought it had broken down completely.[50] However imaginative his description of the early fourteenth century, it probably does represent what he thought coroners were or should be doing in his day. It also sits with the idea that the main purpose of medieval affinities (and of service more generally) was military; raising an army and providing justice were the two main functions of medieval kings.[51] Justiciars and sheriffs had martial roles under the king and it would be surprising if their subordinates in turn did not sometimes participate in these.

Other evidence suggests coroners had multiple roles, a legacy of the office's complex origins and enduring flexibility. The accounts rendered by Patrick Lawmondsone as coroner of Cowal (Argyllshire) between 1445 and 1450 show him collecting revenue from land and wardships and disbursing it to others for the costs of justice and for looking after local defence ('*pro custodia turris de Dunovane*' and '*pro custodia castri de Dunbrettane*').[52] Here the official was responsible for the running of the crown's full rights whereas elsewhere this post is termed a bailie or steward in relation to earldoms or lordships that had come into royal hands. Justice and finance were, of course, inextricable during the Middle Ages, but we find a similar mixing of the functions of steward or mair, judicial officer and peacekeeper in traveller Martin Martin's rehearsal of what the holder, Alexander Fullerton, told him of his role as king's coroner on Arran in the 1690s:

he has his right of late from the Family of *Hamilton*, wherein his Title and Perquisites of Coroner are confirmed to him and his Heirs. He is oblig'd to have three Men to attend him upon all Publick Emergencies, and he is bound by his Office to pursue all Malefactors and to deliver them to the Steward, or in his absence to the next Judge. And if any of the Inhabitants refuse to pay their Rents at the usual term, the Coroner is bound to take him Personally or to seize his Goods. And if it should happen that the Coroner with his retinue of three Men is not sufficient to put his office in execution, then he Summons all the Inhabitants to concur with him; and immediately they rendezvous to the place, where he fixes his Corner's Staff. The Perquisites due to the Coroner are a Firlet [firlot; a quarter of a boll] or Bushel of Oats and a Lamb from every Village on the Isle, both of which are punctually paid him at the ordinary Terms.[53]

Fullerton was a crown coroner, but he effectively worked for the Duke of Hamilton, the hereditary grant of office or the combination of duties in the hands of a single person with subsequent delegation further serving to confuse the issue for historians. A strong arm supplemented Fullerton's halberd and the right to summon numbers in the deployment of physical force pursuant on the administration of justice, reminiscent of a *toíseachdeor*.

At certain junctures, the diverse components of the coroner's job became individually prominent. During the second half of the sixteenth century coroners became important as multi-functional officers of the crown at a period when government was intensifying at all levels. Their role in military organisation is shown when regent Morton issued gunpowder to a coroner at the abbey of Holyrood in 1571, possibly in connection with 'twa greit skirmissings betwixt thame of Leith and Edinburgh; in the quhilkis Mr James Haliburtoun provest of Dundie and crownar to my Lord regentis men of weare [war], wes tane and brocht to Edinburgh castle ...' on 30 August.[54] Morton also gave 6 iron 'graippis [forks] with thair schaftis' to 'my Lord crownar' and paid money to a crowner under the heading of 'men of weir [war]' (both in 1572); three dozen planks went to build a guard house at St Cuthbert's Church (under Edinburgh castle) in April 1573.[55] Another example is 'the ryt honorable' Captain Henry Balfour, described as 'cronar to the scottis companzie men of weir' in Flanders, in a receipt of October 1577 for £687-10-0 Scots issued by John Uddart, an Edinburgh burgess, to Balfour and his captains and lieutenants.[56] These were king's men appointed to supply soldiers, but crowners with delineated territories within Scotland ('of' not 'to') also exercised military functions. The Balfour example shows that coroners were different from captains, as demonstrated again by the description of Archibald Campbell as captain of Dunstaffnage castle and crowner of Lorne (Argyllshire) in 1538.[57] During the sixteenth century, Continental armies gradually acquired more, and more specialised, officers, as part of a movement towards more sophisticated drill and manoeuvre. Abroad once again, Hans Kerensone 'croner' took a receipt from Captain Thomas Muschamp in May 1622. Muschamp was an English officer in Swedish service, serving in a Scottish regiment, and married to a Scot. Kerensone was not a colonel commanding a regiment, a position occupied by someone else. The word 'colonel' arises from Italian '*colonella*', leader of a column. In French and Spanish it became 'coronel', at least for a while, and the Scots seem to have adopted this form of the word on some occasions.[58] The Scottish crowner, in a foreign military context, is more like a steward or supervising quartermaster, charged with getting value for money from military contingents and ensuring their payment and supply. This role does seem to fit with what domestic equivalents sometimes did in the sixteenth century. More usually, however, colonel is quite a different office from crowner: purely military rather than military, judicial and administrative. Thus the two words 'coroner' and 'colonel' are occasionally confused (and confusing) in late-sixteenth- and early-seventeenth-century usage, even if for modern readers they are etymologically quite different.

Coroners' remits could nevertheless be almost boundless. In 1590 the Privy Council ordered 'coronellis' of Dumfries, Eskdale and other parts, along with their deputies, to take action against 'maisterles and unanswerable personis': those without a lord to supervise and be accountable for them.[59] In 1606 when James VI placed troops at Dunyvaig castle on Islay, its owner Angus McConeill or McConnell and three 'crunairs' or 'coronells' of the 'Oo and clergy', 'Harrie' and Rhinns of Islay were bound to pay for their maintenance.[60] Principal Robert Baillie reported that troops raised for Ayrshire in 1639 were 'under my Lord Lowdoun's conduct as crouner' and of the army so raised: 'Our crowners for the most part were noblemen'.[61] When advocate Sir Thomas Hope of Craighall pursued two alleged murderers in 1642 he asserted that their act would have been so counted even if committed 'under ane martiall government of generall or crowneris'.[62] Here the word 'colonel' comes out of 'coroner' and other examples of variant spellings of both words in the *Dictionary of the Older Scottish Tongue* make the connection clear.[63] Hope's comment may have been an echo of debates over 'marshal law' that contemporary

DOI: 10.1057/9781137381071.0008

English parliaments had criticised, this coming out of extensions of the work of the Tudor and early-Stuart provost-marshals 'in the sense of a summary process applied by soldiers to the civilian population in times of urgency'.[64]

Again there may be similarity with *toíseachdeors* for the military functions of sixteenth-century Scottish coroners (and, as we shall see, some of those for the north of England) are reminiscent of twelfth- and thirteenth-century English, Welsh and Irish serjeants or keepers of the peace – known as *ceisiadau* in Welsh[65] – 'appointed as occasion demanded to aid the sheriff in the policing of the shire. They supervised the assessment to arms and arraying of the shire levies, acting as their captains in operations against internal and external enemies'. During the fourteenth century, keepers became 'Justices of the Peace' with purely judicial roles.[66] Originally appointed only to specific localities, keepers became generalised after the English baronial revolt of 1263–5.[67] Late-thirteenth- and early-fourteenth-century English keepers mainly concerned themselves with policing their shire: they restored plundered goods and protected the church; proclaimed commissioners of the peace; took surety for prosecution; attached or summoned the accused; empanelled juries; led shire levies.[68] To underline their military origins, they were sometimes called *capitanei et custodes pacis* (leaders and keepers of the peace). Historian William Rees tried to draw a distinction between what he called 'police serjeants' and 'the later *military* serjeants and the Keepers of the Peace', yet such a distinction is hard to sustain in practice for the twelfth and thirteenth centuries.[69] Military functions were, unsurprisingly, prominent in the northern counties, where a commission of 1307 appointed *custodes pacis* (custodians of the peace) 'for the better preservation of those counties [Cumberland, Westmorland, Northumberland, Lancashire] from incursions of the king's enemies and to punish rebels'.[70] Beyond this keepers could not go, and what distinguished them from later Justices of the Peace was the inability to decide on the offences presented or bills submitted before them.[71] The wardens of the Scottish and Welsh marches were, arguably, the most enduring examples of keepers of the peace.[72]

Keepers might originally have had judicial roles and they certainly developed them during the fourteenth century, but in their twelfth- and thirteenth-century form, they were broadly pro-active in the administration of justice. This remained the case in Ireland for much longer and, as Robin Frame puts it, 'the mantle of the English justices of the peace sat uneasily on the shoulders of the Irish keeper, who was primarily an arrayer and captain in march warfare.[73] In Ireland, keepers were still 'engaged in forcible peace-keeping' until the sixteenth century and such functions were widely dispersed, not only among sheriffs in royal counties and seneschals elsewhere, but also urban magistrates and even bishops.[74] English institutions transferred imperfectly to Ireland, even at this period, and the administrative situation was fluid, reversible and hybridised, so that activities by a keeper or coroner might be taken up by other officials including the constable, captain and even the new president and council in the provinces. There were still chief serjeants of fee in sixteenth-century Ireland and all Irish serjeants seem to have chosen sub-serjeants to do much of their work of serving and executing writs.[75] Ireland had seneschals in the Middle Ages, but a new version came in the 1540s, resembling not only Scottish coroners at certain times and places, but also medieval *toíseachdeors* in having powers to collect rents and dues from Gaels and bring them under law, backed up by soldiers and sometimes martial commissions in areas like Co. Wicklow.[76] Some early modern authorities thought the word seneschal combined the 'ancient' words for justice and officer or governor, though at this level of administration they treated the holder as more like a sheriff.[77] A seneschal presided over the seventeenth-century court of the palatinate of Tipperary; elsewhere a seneschal was like a manorial steward.[78] In Ireland, the offices of seneschal and coroner were sometimes linked and appointments were still made in both to a single individual in the Victorian age.[79]

In Scotland itself, other positions that never had the same presence as in England included local constables, established alongside Justices of Peace under acts of 1587, 1609 and 1617.[80] In England 'Tudor legislation transformed constables incrementally from executive legal officers of the manorial lord to local administrators for the Justices of the Peace', responsible for a range of important tasks not only in law and order, but also taxation.[81] Constables were hard to recruit in Scotland, deployed only intermittently in periods when vagrancy was a serious problem or when the needs of warfare stretched the governmental apparatus.[82] Except for a brief time under Charles I and again during the Cromwellian occupation of the 1650s, Justices only became important after 1708–9 when constables can be found working with them to prepare presentations, executions of brieves or writs, and arrestments (seizures of assets in the hands of a third party) for the Justiciary Circuits in some counties, as well as committing

DOI: 10.1057/9781137381071.0008

offenders to ward – as envisaged at ayres by the act of 1587 and possibly practised at circuit courts during the Interregnum.[83] Constables became much more important in eighteenth-century Scotland. A judgement of the Court of Session (Meldrum v. Brown, 23 December 1746) established that constables could, by their own authority, commit in order to trial as well as executing warrants for Justices. Legal developments after 1748 cemented the position of both officers in maintaining the peace.[84]

In earlier times, coroners did certain similar jobs, carrying on the medieval tradition where judicial functions (including the office of justiciar) were not clearly distinguished from governmental and administrative responsibilities. One of the duties expected of the hereditary coroner of Bute was 'to keep a sufficient Number of Men for apprehending Thieves and Robbers, and detaining them till presented to the Sheriff, and for putting Thiggers [beggars] out of the Isle.'[85] Other officers whose titles might seem ill suited to their allotted tasks found themselves pressed into service at a time of governmental expansion. In 1523, the King's Council ordered stewards, chamberlains and a mair of fee in the east-central Lowlands to provide materials of war to the host at Leith.[86] Legislation of 1545 directed officers at arms, who were strictly executive legal and heraldic officials, to help sheriffs, stewards, and bailies to poind (seize or distrain) the moveable assets of those late with their taxes.[87] In the late sixteenth century the Privy Council even charged the king's almoner, Mr Peter Young of Seton, with supervising actions against sturdy beggars and in 1594 it ordered him to assist the justices at an ayre in Aberdeen; in preparation, 'valentynes [writs or precepts] of maist notabill offendouris be direct to the maisteris, landlordis and baillies, and all given in portuus [porteous roll; the official list of accused persons to be tried] to the crowneris for arresting of thame.'[88] An ambassador to Denmark in the late 1580s was described as the 'Crownar' – probably the Earl Marischal – a reminder that coroners as well as officers at arms might have diplomatic roles.[89] Like many Scottish officials, coroners were subject to a continuing process of adaptation and adjustment to meet the challenge of the rapidly changing political, judicial, administrative and social environment, notably under James VI.

The adaptability of Scottish coroners in the sixteenth century invites us to consider that, as with *toíseachdeors* in earlier times, it may be anachronistic to expect royal or noble appointees to have had clear and consistent job descriptions. For one thing, duties and powers of offices were seldom set out with any precision, but developed over time by

DOI: 10.1057/9781137381071.0008

balancing aspiration with reality in a spirit of compromise. Like any servants of important people, crowners also worked within an open-ended understanding of service, which meant they could legitimately be asked to do anything not incompatible with their honour, in what was a decidedly non-bureaucratic polity.[90] The king's business or 'erands', after all, encompassed almost everything in the realm.[91] At the same time, what a holder made of a job depended on his personality and that of his immediate superior, as well as the circumstances in which both operated.[92]

Broad understandings of patronage and clientage prevailed in other contexts. Tacksmen or landholders with formal leases, usually cadets of the main landowning family in the seventeenth-century Highlands and Islands, were, for example, expected to be loyal administrative and military officers for their lord as well as payers of rents and services.[93] Furthermore historian Simon Kingston reminds us of the 'bewilderingly ill-defined range of names...caught between the usages of two codified systems of hierarchy...Names which had once had precise brehon law significance continued to be used but were no longer prescriptive of offices with exclusive rights and duties.'[94] Over time, the office of *toiseach-deor* acquired a more clearly defined legal meaning: 'basically sheriffs' according to Kingston, though perhaps coroner is closer?[95] Yet once again, contemporaries were sometimes imprecise with the labels they used, perhaps not always willing or able to differentiate between offices and officials.

Meanwhile, for centuries Scottish government 'often worked with the grain of other sources of authority and governance'.[96] Seldom either specified or challenged, the exact balance between noble and royal power remains elusive to the historian. For example, coroners and their serjeants in fourteenth-century Dumfriesshire had to take account of baronial jurisdiction and privilege, even if we do not know what this meant in practice.[97] The character of the coroner and the people he had to work with (and against) were likely crucial to the functioning of the office in different localities and over time. The interchangeable or varied references to different officers demonstrate too that medieval Scotland did not possess a uniform administrative structure. Yet the wide range of jobs assigned to people sporting the title of coroner, from high politics to revenue collection, shows the potential importance of the office.

Flexibility and imprecision is important because, as with Sir John Skene, using a title argued for continuity and uniformity as much as

DOI: 10.1057/9781137381071.0008

it described actual historic roots. Later jurists like Skene sought to create the appearance of a coherent lineage. They ended up creating confusion by conflating offices which sometimes show similarities, but more usually had quite different origins and functions. In 1539 Alane M'Lane became *toschachdoir* of Kintyre (Mull to Altasynach) and in May 1550 Colin Campbell of Ardkinglas became 'officio coronatoris, alias *thochisdoir* de Cowale', suggesting either that the crown conceived the offices as synonymous or wanted them to be treated in that way, investing Gaelic words with new Scots-law meaning.[98] This carried on a process of integrating foreign and native cultures, shown in the assize of William I and still evident in the fourteenth century when, for example, Patrick Lindsay of Bonhill became *toíseachdeor* and forester of Lennox.[99] The conflation was transitional as the crown drew Gaeldom under its control and nineteenth-century historian William Forbes Skene found *toíseachdeors* mostly mentioned in the west – Cowal, Craignish, Kintyre, Knapdale, Lennox, Lochaber and Nithsdale – though he also noted one in Aberdeenshire.[100] Between 1477 and 1546, for example, Strathdoun changed from a 'tochdoreship' to a 'bailliarye' (which may explain Sir John Skene's interpretation).[101] Linking the titles still happened in some much later examples. Documents describe the small estate of Crannich on north Lochtayside, conveyed to Sir Duncan Campbell of Glenorchy by Alexander Menzies of Menzies in 1596, as a thanage (with an associated 'toschadoraschip' of Argewane) in the barony of Weem until at least 1642; Kintyre remained an all-encompassing 'heretable Crownership and Toshdorich or Majorship [mairship]' even in 1685.[102]

As suggested when questioning Sir John Skene's interpretation of the office, the coroner encountered in Scotland from the fourteenth century looks much more like a *toíseachdeor* than his English namesake. Or a superior sheriff's mair, a traditional office denigrated (with feeling) by Skene as occupied by men who 'knawis nocht their office, but ar idle persones, and onely dois diligence in taking up of their fees, from them to quhom they do na gude, nor service to the King'.[103] Again, this was argumentative for mairs and serjeants already did jobs that he thought more suited to coroners. Some mairs were also hereditary holders, known strictly as 'mairs of fee', with weighty offices in, for example, twelfth- and thirteenth-century Galloway and Fife.[104] Later mairs were like sheriffs' officers.[105] That the mair was not exactly the same as the coroner is clear when the two words were used in describing fees payable to a single holder at Renfrew in 1494, or the holding in 1553 by William,

Lord Sempill and thereafter by his son Robert, of the 'crownarschip and mairschip of Blakcairt and Laverane'.[106] The basis for fees payable to the mair of the lordship and regality of Kilbride in the mid-seventeenth century was landholding, whereas the coroner's fees came from a levy on those amerced in court.[107] A regality was like a barony, an area where public jurisdiction was in private hands, though its jurisdiction was much more extensive; areas without regality jurisdiction were known as 'royalty'. The *Regiam Majestatem* also suggests a distinction when setting out the form of citation requiring officers to state their authority: 'si fuerit Marus Domini Regis, vel Toscheoderach ipsius' (if as our Lord king's mair, or his *toíseachdeor*).[108] And again when John MacLachlan of Strathlachlan granted Alan, son of John Riabhach MacLachlan of Dunadd, the offices of 'seneschall' and 'thoisseachdeowra' of lands at 'Glassry' (Glassary, Argyllshire) in the barony of that name in 1436; in contrast, early-sixteenth-century holders received sasine (English 'seisin') as seneschal, but not *toíseachdeor*.[109]

Whatever the early equation of offices, the evidence suggests a separation of the roles of sheriff and coroner no later than the fourteenth century. Conflation or amalgamation may be indicated by the case of Sir Walter Moigne, sheriff and coroner of Aberdeenshire from 1361, and sometimes this linking makes the offices look interchangeable.[110] It is unclear if dual holders actually performed both roles (a topic discussed later) or were merely responsible for ensuring performance. Most documents of the fifteenth and sixteenth centuries suggest that the sheriff and coroner were different, as for example in 1404 when Robert de Halket, sheriff of Kinross, was infeft in the lands of Ballingal, and in his office of 'Coronership of the Shyre of Kinross'; an Exchequer roll for 1489, that includes arrears accounted for by the deputies of Patrick, Earl of Bothwell, as sheriff of Edinburgh for listed goods and plenishings (household furnishings) 'per coronatorem intromissis' (dealt with or sent in by the coroner); and charters of 1528, 1547, 1574 and 1625 to Earls of Buchan, who held both offices separately.[111] By this time, sheriffs rather than coroners were supposed to collect forfeitures, even if coroners did the leg work. The Jedburgh ayre of 31 October 1502 records '*bona arrestata per Coronatores de Jedworth*', the goods specified as the 'oxin, ky, hors, scheipe, ruks of corne, nolt' and so on, belonging to named individuals.[112] Not long after this, in May 1508, Thomas Kennedy of Dunrod, principal coroner of Carrick, signed off fines due to the crown by John Schaw of Hale after the recent justice-ayre at Ayr, because Schaw had withheld payment, allegedly through fear of George Campbell, sheriff-depute of Ayr.[113]

Coroners nevertheless remained under the control of justiciars until at least the sixteenth century. They were executive legal officers or bailiffs, who operated at justice-ayres and received securities from litigants, arrested (cited) indicted criminals, enforced attendance at hearings, or seized forfeited goods in cases involving breach of the king's peace (the sheriff himself was responsible where no such claim was made).[114] In addition to the pleas of the crown the justice-ayres had civil jurisdiction over breaches of the rules of landownership and also heard appeals from the Sheriff Courts and pleas concerning more than one sheriffdom. Ayres thus dealt with the weightiest and often most difficult judicial matters in the land. A mandate signed by James IV and dated 12 April 1504 'commands Maister Richard Lawson of ye Herrige [High Rigs; the Justice Clerk], till tak sourete [surety] of William Dudyngston, crownar of Twedaile, under ye paine of doubilling of the unlaw [fine] of tene punde. That the said Crownar shall enter Jok Wilson's dwelland in the Glene upon Tueyd, in ye next Justice-Ayre of Peblis that sal be halding, sic lik as the said Crownar suld haf enterit him in the last ayre'. It went on to instruct Lawson to write to the sheriff of Tweeddale and his deputes to cease poinding the crowner or his goods for the unlaw hanging over from the last ayre.[115] Here both sheriff and coroner were under the justiciar, though the sheriff was ultimately responsible for the appearance of the accused in court. With the growing importance of sheriffs in the fifteenth and sixteenth centuries coroners came increasingly within their *de facto* control rather than the justiciar's. Legislation in 1487 envisaged crowners being dependent on sheriffs for their share of forfeited goods – like the unshod 'dantit' (broken) horses of convicted murderers awarded by an earlier law.[116] Meanwhile in 1557 the queen delegated the sheriff of Perth to pronounce on conflicting claims to the crownership of Strathearn.[117]

References to late-medieval coroners relate primarily to lands in the south-west and the western Highlands and Islands assimilated by the Scottish crown relatively recently. There was a royal coroner, even in an apparently independent province like Annandale (a stewartry or 'royal regality', not re-absorbed into the royalty of the kingdom) which excluded other royal officers, whom the king could select and instruct to arrest and prosecute serious crimes before royal justices.[118] This may show that royal control over justice in regalities was closer than suggested by their extensive privileges (removal of jurisdiction over the pleas of the crown from the royalty). Perhaps some regalities were more regal than others. The king was, after all, the ultimate guarantor of justice and had, since the time of David I, asserted that royal justice should take effect if the lord's failed.[119] Whether he could enforce that claim was, of course, another matter and only after 1493 were serious attempts made to introduce justiciars and sheriffs into 'the north Ilis and south Ilis' (the Lordship of the Isles); legislation of 1504 made Inverness or Dingwall the seat of justice for the north and Tarbert or Campbeltown for the south.[120] Long after this, the sheriff of Inverness seems largely to have ignored the northern and western parts of his jurisdiction.[121] Coroners were not, however, exclusively western. The examples just given, from Border shires like Berwick, Roxburgh and Peebles, show that this region too presented problems of law, order and government until the seventeenth century, which coroners helped solve.[122] The office protected certain royal interests, particularly important in such core areas, which had, after all, to be governed as well, if not better, than peripheries; feuds that occurred in the Central Lowlands and Southern Uplands were of most concern to late-medieval and early modern Scottish government.

Indeed the office was far from absent from areas close to the centres of power. There are equally frequent mentions of coroners in Banffshire, Carrick, Cunningham, Dumbartonshire, Lanarkshire, Perthshire, Renfrew and the regality of St Andrews. In David II's reign Adam Coussor held the 'office of Cronarie' in Berwickshire; his equivalent in Forfar and Kincardine was Alexander Strathquin; Fife and Forfar belonged to Allan Erskine, Dumfries to Thomas Durance.[123] As far as royal administration was concerned, many of these too had been 'frontier zones' in the twelfth century because they were 'beyond the reach of sheriffdoms, burghs and justices'; some remained so much longer.[124] Operating areas varied considerably and perhaps also the powers between minor and major jurisdictions. The Glenorchy example cited earlier was rather small, whereas some coronerships embraced more than one shire. Coronerships pertained prominently to sheriffdoms, but can also be found in administrative districts like stewartries (Annandale, for example), seneschalships (Kirkcudbright) or regalities (Garioch). There could be coroners without discrete sheriffdoms in the same locale, as in the case of Caithness (not separated from Inverness until 1503) or Sutherland (1633).[125]

In the late Middle Ages Scottish coroners were important officials who helped to consolidate royal power through the implementation of justice. Their story reflects both the strengths and weaknesses of the Scottish crown when it delegated jurisdiction to others, especially the enduring need to rely on local power bases to govern and police. Behind the development and eventual decline of the office during the later sixteenth and seventeenth centuries lies the regularisation of state power, the emergence of Edinburgh-based courts with new ways of conducting justice and the gradual circumvention of lordship and personal bonds by institutionalised mechanisms for managing social relationships. A cluster of changes, particularly during the reign of James VI, altered fundamentally the political, fiscal and judicial landscape. For example, Privy Council business expanded dramatically in the 1580s and 1590s; Exchequer became a permanent court in 1584, involving a reorganisation of royal finances, and taxation became increasingly regular at the same time. After the 'act anent removing and extinguishing of deidlie feuds' of 1598, bloodfeud declined rapidly (almost disappearing from the Lowlands by c.1640) and so too did the numbers entering into bonds of manrent (service).[126] In a new administrative, social, and legal environment coroners became increasingly anachronistic until, by the early eighteenth century, they were largely obsolete.

Notes

1 *Ancient Scottish poems. Published from the manuscripts of George Bannatyne* (Edinburgh: A. Murray and J. Cochran, 1770), 113. J. Pinkerton (ed.), *Scottish poems, reprinted from scarce editions* 3 vols (London: J. Nichols, 1792), vol. 1, 24.

2 C. Innes, *Lectures on Scotch legal antiquities* (Edinburgh: Edmonston and Douglas, 1872), 84. Advocates were the élite of the Scottish legal profession, the equivalent of English barristers.

3 J. Eaton, 'The coroner's court from the medical standpoint. Part II', *The Provincial Medical Journal* 7 (1888), 345. This is a direct lift from Chambers Encyclopaedia. An article on the Scottish coroner in *The Scotsman* (18 September 1893), p. 9, is in turn cribbed from Eaton.

4 Walker, *Legal history of Scotland*, vol. 1, 217.

5 Wellington, *King's coroner*, 42–9. J. C. Davies, 'Felony in Edwardian Wales', *The Transactions of the Honourable Society of Cymmrodorion* (1916–17), 169–74.

6 'Ordonnance ... sur le gouvernement d'Ecosse', APS I, 121. ODNB 'Boyd family'. *Register of the great seal of Scotland* [RMS] VIII, no. 745. The original document dates Lekprevik's appointment to 1397, but 1307 is a more likely year. J. Bain (ed.), *Calendar of documents relating to Scotland ... vol. II,*

DOI: 10.1057/9781137381071.0008

1272–1307 (Edinburgh: HMSO, 1884), nos 1691 (4), 1909. J. I. Smith, 'Criminal procedure', in *An introduction to Scottish legal history* Stair Society vol. 20 (Edinburgh, 1958), 427, says the office is found as early as the eleventh century, but cites no authority.

7   A. A. M. Duncan (ed.), *The acts of Robert I, king of Scots, 1306–1329* (Edinburgh: Edinburgh UP, 1988), 30–1, 437–8. A. A. M. Duncan, *Scotland: the making of the kingdom* (Edinburgh: Oliver & Boyd, 1975), 496, clarifies the charter, pointing out that the aim was to exclude the sheriff from the town. See H. L. MacQueen, *Common law and feudal society in medieval Scotland* (Edinburgh: Edinburgh UP, 1993), 62–4, on the elevated standing of justiciars.

8   P. G. B. McNeill (ed.), *The practicks of Sir James Balfour of Pittendreich, reproduced from the printed edition of 1754* 2 vols. (Edinburgh, 1962–3), 623–4. P. J. Hamilton-Grierson (ed.), *Habakkuk Bisset's Rolment of Courtis* 3 vols. (Edinburgh: Scottish Text Society, 1920–6), vol. 2, 212–13.

9   J. Skene, *De Verborum Significatione* (Edinburgh: Robert Walde-grave, 1597).

10 A. A. M. Duncan, '*Regiam Majestatem*: a reconsideration', *Juridical Review* 73 (1961), 199–217.

11 Quoted in Goodare, *Government of Scotland*, 260. Goodare is quoting Balfour's paraphrase (p. 510) of the *Regiam Majestatem* IV.8.3, which states that the woman should go '*ad capitalem marum illius comitatus, vel ad Toscheoderach, si poterit inveniri*'. Smyth, *De republica Anglorum*, 2:21 (p. 73), disposed of the idea that the English word coroner comes from the Latin for a public hearing.

12 J. D. Ford, *Law and opinion in Scotland during the seventeenth century* (Oxford: Hart, 2007), 57. We might note that the law of rape required penetration and emission in England, but not in Scotland.

13 Stewart-Brown, *Serjeants of the peace*, 5.

14 *Ibid.*, 117–18. In the fourteenth century, the serjeant's helpers were two under-serjeants. A. M. Tonkinson, *Macclesfield in the later fourteenth century: communities of town and forest* (Manchester: Carnegie, 1999), 44–5, 144.

15 Stewart-Brown, *Serjeants of the peace*, 5–8, 73–86. W. Rees, 'Survivals of ancient Celtic custom in medieval England', in H. Lewis (ed.), *Angles and Britons: O'Donnell lectures* (Cardiff: University of Wales Press, 1963), 155–7.

16 F. W. Maitland, *Select pleas in manorial and other seignorial courts* volume 1 (London: Selden Society, 1889), xxx, xxxiii. W. A. Morris, *The frankpledge system* (London: Longmans, Green, 1910), 43, 45–59. Stewart-Brown, *Serjeants of the peace*, 99–104. D. A. Crowley, 'The later history of frankpledge', *Bulletin of the Institute of Historical Research* 48 (1975), 1–15. H. Summerson, 'Peacekeepers and lawbreakers in medieval Northumberland, c.1200–c.1500', in M. Prestwich (ed.), *Liberties and identities in the medieval British Isles* (Woodbridge: Boydell, 2008), 58–9, 61.

DOI: 10.1057/9781137381071.0008

17 C. Schmid, 'Border lordship and central politics: the local context of Scottish power struggles, 1525–1552' (Guelph University MA thesis, 2007), 40–51. Hunnisett, *Medieval coroner*, 27–8.

18 F. A. Bailey (ed.), *Selection from the Prescot court leet and other records, 1447–1600* (Manchester: Records Society of Lancashire and Cheshire, 1937), 191.

19 NA STAC 8/183/51. S.P. 14/103/68 and 14/109/6. *Acts of the Privy Council of England. New series [vol.] 35, 1616–1617* (London: HMSO, 1927), 380–3. S. J. Watts, 'Tenant-right in early seventeenth-century Northumberland', *Northern History* 6 (1971), 73–4.

20 See for example W. Robertson, *An index, drawn up about the year 1629: of many records of charters, granted by the different sovereigns of Scotland between the years 1309 and 1413, most of which records have been long missing* (Edinburgh: Murray & Cochrane, 1798), 'Coroner, office of'.

21 APS I, 58. Walker, *Legal history of Scotland* vol. 1, 217.

22 M. Bateson (ed.), 'The Scottish king's household and other fragments from a 14th century manuscript', *Miscellany of the Scottish History Society II* (Edinburgh: Scottish History Society, 1904), 36–7. G. W. S. Barrow and W. W. Scott (eds), *The acts of William I king of Scots, 1165–1214* (Edinburgh: Edinburgh UP, 1971), 46 and no. 80. E. L. G. Stones (ed.), *Anglo-Scottish relations, 1174–1328: some selected documents* (London: Nelson, 1965), 123–4. Hume, *Commentaries*, vol. 2, 24n. F. J. Watson, *Under the hammer: Edward I and Scotland, 1286–1306* (East Linton: Tuckwell, 1998), 217.

23 Stewart-Brown, *Serjeants of the peace*, 64–5. Hunnisett, 'Origins', 92–5, 98–9. Hunnisett, *Medieval coroner*, 34. H. G. Richardson and G. O. Sayles, *The governance of mediaeval England from the conquest to Magna Carta* (Edinburgh: Edinburgh UP, 1963), 186–8.

24 J. Campbell, *The Anglo-Saxon state* (London: Hambledon, 2000), 52.

25 P. N. R. Zutshi (ed.), *The letters of Frederic William Maitland, vol. 2* (London: Selden Society, 1995), no. 175.

26 W. C. Dickinson, *Scotland from the earliest times to 1603* (1961. 3rd edition revised and edited by A. A. M. Duncan. Oxford: Oxford UP, 1977), 96–8. H. L. MacQueen, 'Scots law under Alexander III', in N. H. Reid (ed.), *Scotland in the reign of Alexander III, 1249–1286* (Edinburgh: John Donald, 1990), 82.

27 A. J. S. Gibson and T. C. Smout, *Prices, food, and wages in Scotland, 1550–1780* (Cambridge: Cambridge UP, 1994), 365–75. R. D. Connor and A. D. C. Simpson, *Weights and measures in Scotland: a European perspective* (East Linton: Tuckwell, 2004).

28 Walker, *Legal history of Scotland*, vol. 2, 337–40. J. W. Cairns, 'Historical introduction', in K. Reid and R. Zimmermann (eds), *A history of private law in Scotland. Volume 1: introduction and property* (Oxford: Oxford UP, 2000), 14–184. C. A. Malcolm, 'The office of sheriff in Scotland: its origin and early development', *Scottish Historical Review* 20 (1923), 135. RPCS IV, 218. J. Cameron

DOI: 10.1057/9781137381071.0008

(ed.), *The justiciary records of Argyll and the Isles, 1664–1705, vol. 1* Stair Society 12 (Edinburgh, 1949), xx. G. Hutcheson, *Treatise on the offices of the justice of peace; constable; commissioner of supply; and commissioner under comprehending acts in Scotland* 3 vols (Edinburgh: W. Creech, 1806–8), vol. 1, 4–5. I am grateful to Alex Woolf, John Finlay and Athol Murray for advice on this topic.

29 K. Simms, *From kings to warlords: the changing political structure of Gaelic Ireland in the later Middle Ages* (Woodbridge: Boydell, 1987), 11.

30 A. C. Lawrie (ed.), *Early Scottish charters prior to A.D. 1153* (Glasgow: James MacLehose, 1905), 220, 347, speculated that toíseach, in the notes to the Book of Deer, was a term applied, by someone of Irish origin, to a native office actually called something else. An exhaustive historiographical survey of understandings of the word can be found in D. Broun, 'The property records in the Book of Deer as a source for early Scottish society', in K. Forsyth (ed.), *Studies on the Book of Deer* (Dublin: Four Courts, 2008), 315–56. See also A. O. Anderson, *Early sources of Scottish history, A.D. 500 to 1286* 2 vols. (1922. Stamford: Paul Watkins, 1990), vol. 2, 174–81. A. Woolf, *From Pictland to Alba, 789–1070* (Edinburgh: Edinburgh UP, 2007), 342–50. G. Márkus, 'Dewars and relics in Scotland: some clarifications and questions', *Innes Review* 60 (2009), 95–144. The present book was largely written before the appearance of Márkus's valuable piece.

31 Dickinson, *Scotland from the earliest times*, 51, 53–5. W. F. Skene (ed.), *John of Fordun's Chronicle of the Scottish nation* 2 vols (Edinburgh: Edmonston & Douglas, 1872), vol. 2, 459, suggests that 'dior' is 'an old word, signifying "of or belonging to law"'. I am grateful to my colleague Alex Woolf for guiding me through the terms.

32 A. Grant, 'Thanes and thanages, from the eleventh to the fourteenth centuries', in A. Grant and K. J. Stringer (eds), *Medieval Scotland: crown, lordship and community* (Edinburgh: Edinburgh UP, 1993), 42.

33 J. Stuart (ed.), *The Book of Deer* (Edinburgh, 1869), lxxxi. Innes, *Scotch legal antiquities*, 79–80, 84. J. Cameron, *Celtic law* (Edinburgh: William Hodge, 1937), 233. K. A. Steer and J. W. M. Bannerman, *Late medieval monumental sculpture in the west Highlands* (Edinburgh: HMSO, 1977), 143. R. A. Dodgshon, *Land and society in early Scotland* (Oxford: Oxford UP, 1981), 62. Woolf, *Pictland to Alba*, 342–4. A. Grant, 'Franchises north of the border: baronies and regalities in medieval Scotland', in M. Prestwich (ed.), *Liberties and identities in the medieval British Isles* (Woodbridge: Boydell, 2008), 177.

34 Skene (ed.), *Fordun's chronicle*, vol. 2, 459. APS I, 58, 380, 599, 633. Dodgshon, *Land and society*, 65. Serjeants had the right to pursue, attach and indict criminals. W. C. Dickinson (ed.), *The court book of the barony of Carnwath, 1523–1542* Scottish History Society 3rd series 29 (Edinburgh, 1937), lxxxv–vi. W. C. Dickinson, 'Surdit de sergaunt', *Scottish Historical Review* 39 (1960), 170–5. A. B. Calderwood (ed.), *Acts of the lords of council* vol. 3 (Edinburgh:

DOI: 10.1057/9781137381071.0008

HMSO, 1993), 138, 227. The term serjeant was mostly used in Galloway – and in Wales and the north of England. H. L. MacQueen, 'The laws of Galloway: a preliminary survey', in R. D. Oram and G. P. Stell (eds), *Galloway: land and lordship* (Edinburgh: Scottish Society for Northern Studies, 1991), 131–43. A. Grant, 'The construction of the early Scottish state', in J. R. Maddicott and D. M. Palliser (eds), *The medieval state* (London: Hambledon, 2000), 55, sees the officer being in charge of a holy object used in court for oath-swearing. Márkus, 'Dewars and relics', 98–102, questions this association, saying the keeper of relics is a different official.

35 *Balfour's Practicks*, 566. MacQueen, 'Laws of Galloway', 133–4. M. R. Gunn, *History of the clan Gunn* (Glasgow: A. MacLaren, 1969), 44. Grant, 'Franchises north of the border', 179–80. *The acts of the lords auditors of causes and complaints, 1466–1494* (London: House of Commons, 1839), 100. Innes, *Scotch legal antiquities*, 69–72. Early modern Shetland had 'ranselmen' who were responsible for investigating theft and keeping the peace. G. Goudie, 'The ancient local government of the Shetland Islands', *Universitets Jubilæets Danske Samfunds* 4 (1886), 276. G. Goudie, *The Celtic and Scandinavian antiquities of Shetland* (Edinburgh: W. Blackwood, 1904), 241–6. Commissions in any lawless zone allowed independent action by a number of officials. R. B. Armstrong, *The history of Liddesdale, Eskdale, Ewesdale, Wauchopedale and the debateable land* (Edinburgh: D. Douglas, 1883), 10.

36 Márkus, 'Dewars and relics', 100.

37 NAS GD112/39/29/1 (1619).

38 I owe this point to Dr Alexander Grant.

39 MacQueen, 'Laws of Galloway', 135. Broun, 'Property records', 353–4, sees the earlier *mormaer* and *toíseach* controlling 'general dues and services in their own right: as leading lords, not as officials' of the crown. Simms, *Kings to warlords*, 83.

40 Pollock and Maitland, *English law*, vol. 1, 200.

41 MacQueen, 'Laws of Galloway', 136. Grant, 'Franchises north of the border', 172. K. Stringer, 'States, liberties and communities in medieval Britain and Ireland (c.1100–1400)', in M. Prestwich (ed.), *Liberties and identities in the medieval British Isles* (Woodbridge: Boydell, 2008), 31. M. J. Gunn, 'The clan Gunn and the office of crowner of Caithness', *Clan Gunn Society Magazine* 16 (1980–1), 33–9.

42 *La recherche des singvlaritez plvs remarquables, concernant l'estat d'Escosse. Vové A tresauguste & tresclemente Princesse Marie Roynne d'Escosse, & doüairiere de France. Par Dauid Chambre Escossois, conseiller en la cour de Parlement à Edinbourg, ville capitale d'Escosse* (Paris: Michel Gadoulleau, 1579), 7r. ODNB. A collection of fourteenth-century laws and ordinances for the government of north Wales mentions '*conseruatores pacis vocati Keys*'. Ellis (ed.), *The record of Caernarvon*, iv, 131.

DOI: 10.1057/9781137381071.0008

43 Skene, *De Verborum Significatione*, 'tocheoderache'.

44 W. F. Skene, *Celtic Scotland: a history of ancient Alban* 3 vols (1876–80. 2nd
    edition. Edinburgh: David Douglas, 1890), vol. 3, 279.

45 Quoted in MacQueen, 'Laws of Galloway', 136. A mair of fee was a hereditary
    officer under the crown.

46 C. Innes, *Scotland in the Middle Ages: sketches of early Scotch history and progress*
    (Edinburgh: Edmonston and Douglas, 1860), 193. D. Campbell, *Reminiscences
    and reflections of an octogenarian highlander* (Inverness: Northern Counties
    Newspaper, 1910), 9, 84. Campbell believed the office was hereditary at
    Glenlyon in the seventeenth and eighteenth centuries. I. F. Grant, *The social
    and economic development of Scotland before 1603* (Edinburgh: Oliver and
    Boyd, 1930), 44. MacQueen, 'Laws of Galloway', 137. Some see late medieval
    toíseachs in the same light. Stuart (ed.), *Book of Deer*, lxxxi.

47 ERS V, 601, 658; VI, lxx. Walker, *Legal history of Scotland* vol. 1, 230.

48 NAS GD437/17 and 96. Mair of fee was never equated with coroner, the
    two offices being described conjointly when one person held both in a
    jurisdiction ('and' not 'or'). For example NAS GD150/2282. The same is true
    of bailie and coroner. NAS GD64/1/3/4. GD64/1/4/16.

49 D. E. R. Watt (ed.), *Scotichronicon* 9 vols (Aberdeen: Aberdeen UP, 1987–98),
    vol. 7, 58–9. Coroners could arrest all indicted people throughout the year, not
    only after the proclamation of an ayre. *Balfour's Practicks*, 566. Eilean Donan is
    an iconic castle in Loch Duich in the western Highlands of Scotland.

50 MacQueen, *Common law*, 54. Walker, *Legal history of Scotland* vol. 2, 299–302.

51 See Murray, 'Administrators of church lands', 32, on the military functions of
    bailies. In thirteenth-century Wales the words steward (*distain*), seneschal
    (*synysgal*) and constable (*cwnstabl*) were all used to describe an honourable
    office, sometimes hereditary, whose primary function was military.
    J. B. Smith (ed.), *Medieval Welsh society: selected essays by T. Jones Pierce*
    (Cardiff: University of Wales Press, 1972), 33–4.

52 *The Exchequer Rolls of Scotland, 1264–1600* 23 vols (Edinburgh: HMGRO,
    1878–1908), V, 201–2, 246, 329–30, 358–9, 413–14. ERS VI, lxx. Late medieval
    Sheriff Court sittings were usually held at the royal castle that was the legal
    centre of the shire. R. Nicholson, *Scotland. The later Middle Ages* (Edinburgh:
    Oliver & Boyd, 1974), 18. Nicholson believes that these men were acting as
    ballivi ad extra and held other posts like coroner or chamberlain alongside.
    Ibid., 380. J. M. Gilbert, *Hunting and hunting reserves in medieval Scotland*
    (Edinburgh: John Donald, 1979), 147.

53 M. Martin, *A description of the western islands of Scotland* (London: A. Bell,
    1703), 224. The Fullertons had been coroners there since c.1500. NAS GD1/19/5.

54 *A diurnal of remarkable occurents that have passed within the country of
    Scotland...* (Edinburgh: Bannatyne Club, 1833), 246. This same usage as
    commanding officer is found elsewhere in the volume at pp. 55, 57.

DOI: 10.1057/9781137381071.0008

55　ATS XII, 251, 339, 343.

56　NAS RD1/16, f. 293v. The principal had been borrowed 18 months earlier, when in two documents Balfour stated he could not write. RD1/15, ff. 65, 105 (March and April 1576). RD1/15, f. 65 (14 March 1575). Edward I took coroners with him on his expedition into Scotland. Hunnisett, *Medieval coroner*, 168–9. Balfour seems to have died soon after and his widow remarried a Captain John Balfour. NAS GD2/47 (1584).

57　NAS GD112/17/1/1/2.

58　Krigsarkiv Svenska 0035:0418, 19 May 1622. I owe this reference to my colleague Steve Murdoch and the etymological insights to Julian Goodare.

59　RPCS IV, 800. The Privy Council also charged the king's master almoner Peter Young with enforcing an act against beggars in 1593. APS IV, 42–3. Until its abolition in 1708, Scotland's Privy Council possessed executive, legislative and judicial functions.

60　RPCS VII, 627. In 1591, the king entered into a surrender and regranting with McConnell of extensive holdings in Kintyre amounting to more than 177 merklands given up and 160 acquired 'for service of ward and releif'. NAS RD1/37, ff. 343–5.

61　D. Laing (ed.), *The letters and journals of Robert Baillie, A.M., principal of the university of Glasgow, 1637–1662* 3 vols (Edinburgh: Bannatyne Club, 1841–2), vol. 1, 200–1, 211.

62　J. I. Smith (ed.), *Selected justiciary cases, 1624–50* (Edinburgh: Stair Society, 1972), 488. This seems to follow J. Achesone, *The military garden, or instructions for all young souldiers and such who are disposed to learne, and have knowledge of the militarie discipline* (Edinburgh: John Wreittoun, 1629), 25.

63　Pinkerton, *History*, vol. 1, 387. For a comparable military usage from 1645 see J. Maidment (ed.), *The chronicle of Perth; a register of remarkable occurrences chiefly connected with that city, from the year 1210 to 1668* (Edinburgh: Maitland Club, 1831), 42.

64　Harding, *English law*, 69.

65　J. E. Lloyd, *A history of Carmarthenshire* 2 vols (Cardiff: W. Lewis, 1935–9), vol. 1, 218. Rees, 'Survivals of ancient Celtic custom', 155–6. Rees, *South Wales and the march*, 59, 69, 103–7, 244. These may have been more like constables than coroners.

66　R. Frame, *Ireland and Britain, 1170–1450* (London: Hambledon, 1998), 301. Otway-Ruthven, 'Anglo-Irish shire government', 139–40. A. Harding, 'The origins and early history of the keepers of the peace', *Transactions of the Royal Historical Society* 5th series 10 (1960), 85–109. Stewart-Brown, *Serjeants of the peace*, 36–9, 46. J. P. Dawson, *A history of lay judges* (Cambridge, Mass.: Harvard UP, 1960), 136–45. G. Jacob, *Lex constitutionis: or, the gentleman's law* (London: E. Nutt and H. Gosling, 1719), 337.

67　Harding, *English law*, 69. Bellamy, *Crime and public order*, 94–5.

DOI: 10.1057/9781137381071.0008

68   Harding, 'Keepers of the peace', 91–2. Lambarde, *Eirenarcha*, 17–18.

69   Rees, 'Survivals of ancient Celtic custom', 164.

70   Harding, 'Keepers of the peace', 101–2, 104.

71   B. H. Putnam, 'The transformation of keepers of the peace into justices of the peace, 1327–80', *Transactions of the Royal Historical Society* 4th series 12 (1929), 19–48. B. H. Putnam, *Kent keepers of the peace, 1316–1317* (London: Headley Brothers, 1933).

72   Harding, *English law*, 156.

73   R. Frame, 'The judicial powers of the medieval Irish keepers of the peace', *Irish Jurist* n.s. 2 (1967), 326.

74   Frame, *Ireland and Britain*, 233–4, 302–4. As with Scottish coroners, keepers were charged with receiving forfeitures, arresting offenders and taking indictments and sureties. *Ibid.*, 309.

75   Ellis, *Reform and revival*, 201–2.

76   J. G. Crawford, *Anglicizing the government of Ireland: the Irish privy council and the expansion of Tudor rule, 1556–1578* (Dublin: Irish Academic Press, 1993), 206, 278, 416, 447. Maginn, 'Elizabethan Cavan', 79–80. Elizabeth I wanted sheriffs and justices to replace captains and seneschals, but this aim was not always realised.

77   Nelson, *Lex maneriorum*, 168. Public Record Office of Northern Ireland DIo4/5/3/43 (1691). Garnham, *Courts in Ireland*, 74–5. J. Kitchin, *Le court leete, et court baron, collect per Iohn Kytchin de Greys Inne vn appre[n]tice en le ley, et les cases et matters necessaries pur seneschals de ceux courts a scier, pur les students de les measons de chauncerie* (London: R. Tottelli, 1580). T. W. Hall, *South Yorkshire historical sketches* (Sheffield: J. W. Northend, 1931), 16–22, describes the work of William West, seneschal to the Earl of Shrewsbury in Hallamshire and chief steward of the manor of Sheffield, during the 1580s and 1590s.

78   Acts of the Parliaments of Ireland 7 Will. III, c. 19, 'Act for granting tales on tryals, to be had in the court of the country palatine of Tipperary, before the seneschal'. W. Scroggs, *The practice of courts leet and courts baron* (1701. 3rd edition, London: J. Nutt, 1714), 11. E. Bullingbroke, *The duty and authority of Justices of the Peace and parish officers in Ireland* (Dublin: G. Dibson, 1766), 490. Webb, *Municipal government in Ireland*, 196.

79   Public Record Office of Northern Ireland T2429/1, Manor and Lordship of Greencastle and Mourne, Co. Down (1868). An English seneschal was a steward of a franchise. Hall, *South Yorkshire historical sketches*, 16–61. H. M. Cam, *Liberties & communities in medieval England* (Cambridge: Cambridge UP, 1944), 188–90, 194, 217–19.

80   R. Boyd, *The office, powers, and jurisdiction, of his majesty's justices of the peace, and commissioners of supply, for Scotland* (Edinburgh: William Creech, 1794), 1–37. S. Moir, ' "Some godlie, wyse and vertious gentilmen": communities, state formation, and the justices of the peace in Scotland' (University of Guelph Ph.D., 2002).

81   H. R. French, 'Parish government', in S. Doran and N. Jones (eds), *The Elizabethan world* (London: Routledge, 2011), 149. J. R. Kent, *The English village constable, 1580–1642: a social and administrative study* (Oxford: Oxford UP, 1986).

82   G. Tait, *A summary of the powers and duties of a justice of the peace in Scotland* (1815. 2nd edition. Edinburgh, 1816), 69. Whetstone, *Scottish county government*, 27–8, 32–6. J. Innes, 'Governing diverse societies', in P. Langford (ed.), *The eighteenth century* (Oxford: Oxford UP, 2002), 117–18. Davies, 'Scottish legal system', 134. M. Lynch, *The early modern town in Scotland* (London: Croom Helm, 1987), 16, suggests that the office may have been attractive to rising urban craftsmen. There were also more elevated hereditary constables like the Scrymgeours of medieval Dundee. E. P. D. Torrie, *Medieval Dundee: a town and its people* (Dundee: Abertay Historical Society, 1990), 25, 50.

83   See for example NAS JC26/97. *Act of the Justices of the Peace of the Shire of Haddingtoun, for appointing of constables, and more effectually restraining vagrants* (Edinburgh: John Mosman and William Brown, 1729). Moir, 'Justices of the peace', 312–23.

84   James Fergusson of Kilkerran, *Decisions of the Court of Session from 1738–1752* (Edinburgh: Neill & Company, 1775), 304–5. Boyd, *Justices of the peace*, 792–3.

85   W. Forbes, *The duty and powers of justices of peace, in this part of Great-Britain called Scotland* (Edinburgh: A. Anderson, 1707), pt. 2, 11. Thigging usually meant respectable seasonal begging for seed corn, but could extend to all sorts of less genteel mendicancy; here it meant the latter. G. Mackenzie, *Observations on the acts of Parliament…* (Edinburgh: A. Anderson, 1686), 16. J. Grant, *Thoughts on the origin and descent of the Gael* (Edinburgh: Walker and Greig, 1814), 63–4. A. MacDonald, 'Social customs of the Gaels', *Transactions of the Gaelic Society of Inverness* 32 (1924–5), 300–1.

86   R. K. Hannay (ed.), *Acts of the Lords of Council in Public Affairs, 1501–54* (Edinburgh: HMGRO, 1932), 182.

87   APS II, 462–3. They were also involved in 1625. APS V, 185.

88   RPCS first series V, 755. Houston, 'Royal almoner', 29. Porteous rolls originated in the fourteenth century and first survive from the sixteenth century. They contained instructions to various officials to make enquiries whether a variety of offences had been reported and they list the names of those charged. See for example Perth & Kinross Archives B59/26/1/13/10 (1628).

89   Calendar of State Papers, 'the Scottish series, 1509–1603', II, pp. 531, 549, 564; X, no. 116. K. Stevenson, 'Jurisdiction, authority and professionalisation: the officers of arms of late medieval Scotland', in K. Stevenson (ed.), *The herald in late medieval Europe* (Woodbridge: Boydell, 2009), 62–4.

90 R. Horrox, 'Service', in R. Horrox (ed.), *Fifteenth-century attitudes: perceptions of society in late medieval England* (Cambridge: Cambridge UP, 1994), 63.

91 APS II, 296.

92 Murray, 'Administrators of church lands', 38–9.

93 F. J. Shaw, *The northern and western islands of Scotland: their economy and society in the seventeenth century* (Edinburgh: John Donald, 1980), 48–57.

94 S. Kingston, *Ulster and the Isles in the fifteenth century: the lordship of the Clann Domhnaill of Antrim* (Dublin: Four Courts, 2004), 157.

95 *Ibid.*, 158.

96 Stringer, 'States, liberties and communities', 15.

97 Innes, *Scotch legal antiquities*, 69–72.

98 RPS II, 3098. RMS IV, no. 467. Skene (ed.), *Fordun's chronicle*, vol. 2, 459. J. and R. W. Munro (eds), *Acts of the Lords of the Isles, 1336–1493* Scottish History Society 4th series 22 (Edinburgh, 1986), lii and no. A72.

99 Duncan, *Scotland*, 109–10. C. J. Neville, *Native lordship in medieval Scotland: the earldoms of Strathearn and Lennox, c.1140–1365* (Dublin: Four Courts, 2005), 52, 59, 89. In 1491, the Lords Auditors summoned John, Lord Drummond to show his title to the 'officez of Stewartry crovnarschip & forestary of the lordschip of Straithern'. *Acts of the lords auditors*, 150–1. Malcolm Drummond of Concraig acquired the offices from the tutor to the Earl of Strathearn in 1426. NAS GD160/1/8. His heir surrendered them to the crown in 1473, which granted them to John Drummond of Cargill. GD160/1/14. GD160/2/2. In 1503, John Lindsay of Ballhall was called mair of fee of the earldom of Lennox. Calderwood, *Lords of council*, 273.

100 Skene (ed.), *Fordun's chronicle*, vol. 2, 459. Skene, *Celtic Scotland*, vol. 3, 302. Munro, *Lords of the Isles*, lii. *Atlas of Scottish history to 1707*, 190, shows they are mostly found in a swathe running north-east from Kintyre to Aberdeenshire; although some are first mentioned in the fourteenth century most initial references come from the fifteenth century.

101 ERS XV, 630. Similarly sheriff and forester might be combined in the twelfth and thirteenth centuries, though less so as time when on. Gilbert, *Hunting*, 128–52.

102 NAS GD112/2/124/11. GD112/17/1/1/12. The Menzies seem to have held this since the early fourteenth century. The Mackays of Ugadale held the mairship and coronership of north Kintyre from the Lords of the Isles in the fifteenth and sixteenth centuries. RMS I, appendix 2, 476, 652; III, 2756. Munro, *Lords of the Isles*, lii and nos A59, A72. Kingston, *Ulster and the Isles*, 158–9. Walker, *Legal history of Scotland* vol. 4, 468.

103 Skene, *De Verborum Significatione*, 'marus'. Some later writers thought 'serjands or mairs' were sheriffs' officers. J. J. Darling, *The powers and duties of messengers-at-arms…* (Edinburgh: William Tait, 1840), 1.

104 W. C. Dickinson (ed.), *The sheriff court book of Fife, 1515–1522* Scottish History Society, 3rd series 12 (Edinburgh, 1928), lxii–lxvi, 392–3.

DOI: 10.1057/9781137381071.0008

105 Calderwood, *Lords of council*, 235. Innes, *Scotch legal antiquities*, 78.
106 Dickinson (ed.), *Sheriff court book of Fife*, lxv–vi. *Acts of the lords auditors*, 190, 192. RPS IV, no. 2271.
107 NLS Ch. 7981.
108 T. M. Cooper (ed.), *Regiam Majestatem and Quoniam Attachiamenta based on the text of Sir John Skene* (Edinburgh: Stair Society, 1947), I.6.7.
109 Steer and Bannerman, *Late medieval monumental sculpture*, 143.
110 A. A. M. Duncan, 'The "Laws of Malcolm Mackenneth"', in A. Grant and K. J. Stringer (eds), *Medieval Scotland: crown, lordship and community* (Edinburgh: Edinburgh UP, 1993), 256–7. APS II, 561. RMS IV, no. 1736.
111 Duncan, 'Laws of Malcolm MacKenneth', 266. NAS E28/294A. I owe this reference to Athol Murray. On the importance of 1404 and 1488–9 see MacQueen, *Common law*, 54–5.
112 NAS JC1/1, p. 149. Also quoted in Craig, *Coroner*, 19.
113 NAS GD25/2/23.
114 T. D. Fergus (ed.), *Quoniam Attachiamenta* (Edinburgh: Stair Society, 1996), Ch. 39 [p. 223]. Duncan, 'Laws of Malcolm MacKenneth', 256. Bateson (ed.), 'The Scottish king's household', 36–7. M. Sanderson, *Mary Stewart's people: life in Mary Stewart's Scotland* (Edinburgh: James Thin, 1987), 135–48. Strictly speaking, the sheriff was responsible for the accused appearing in court. Dickinson (ed.), *Sheriff court book of Fife*, xlv–vi. For an example of a coroner arresting a prominent landowner at the justiciar's instruction during the reign of David II see NLS Adv. MS. 16.1.10, fo. 43r. For the office and instruments of the early-seventeenth-century bailiff see 'Account of a journey into Scotland, 1629', *The Scottish Antiquary* 9 (1895), 180.
115 NAS JC1/1, p. 179 [inserted paper]. J. Finlay, *Men of law in pre-reformation Scotland* (East Linton: Tuckwell, 2000), 45, 63, 204.
116 APS II, 177. APS I, 710.
117 NAS GD160/4/14.
118 Barrow, *Kingdom of the Scots*, 127. In a personal communication, Professor Barrow has suggested that there may have been a sort of coroner in Annandale as early as 1165x1173, based on Barrow and Scott (eds), *The acts of William I*, no. 80. Walker, *Legal history of Scotland* vol. 1, 223; vol. 2, 337. Stringer, 'States, liberties and communities', 12. Grant, 'Franchises north of the border', 186. Stewartries and bailieries were regalities and baronies that had fallen into the hands of the king, but which remained distinct entities with their own courts.
119 W. C. Dickinson, 'The administration of justice in medieval Scotland', *Aberdeen University Review* 34 (1951), 339–40.
120 APS II, 249 (1504 c. 3).
121 J. Goodare, *State and society in early modern Scotland* (Oxford: Oxford UP, 1999), 225.

DOI: 10.1057/9781137381071.0008

122 M. Laing, *The history of Scotland, from the union of the crowns ... to the union of the kingdoms* 4 vols (1800. 3rd edition. London: J. Mawman *et al.*, 1819), vol. III, 46–8. T. I. Rae, *The administration of the Scottish frontier, 1513–1603* (Edinburgh: Edinburgh UP, 1966), 133–52. Walker, *Legal history of Scotland* vol. 2, 337. M. B. Wasser, 'Violence and the central criminal courts in Scotland, 1603–1638' (Columbia University Ph.D., 1995), 56–8. M. M. Meikle, *A British frontier? Lairds and gentlemen in the eastern Borders, 1540–1603* (East Linton: Tuckwell, 2004), 54, 63.

123 Robertson, *Missing charters*, 30, 50–1.

124 MacQueen, *Common law*, 41.

125 Walker, *Legal history of Scotland* vol. 4, 463, 467.

126 A. Murray, 'Exchequer, council and session, 1513–1542', in J. H. Williams (ed.), *Stewart style, 1513–1542: essays on the court of James V* (East Linton, 1996), 97–117. APS IV, 158–9. K. M. Brown, *Bloodfeud in Scotland, 1573–1625. Violence, justice and politics in early modern society* (Edinburgh: John Donald, 1986). Goodare, *State and society*, 75–6. Braddick, *State formation*, 355–71.

DOI: 10.1057/9781137381071.0008

4
Scottish Coroners from c.1500 until Their Disappearance in the Eighteenth Century

Abstract: *In their heyday during the fifteenth and sixteenth centuries, Scottish coroners serviced itinerant courts of justice. Thereafter they became sidelined by changes in the constitution of courts and in legal procedures, especially the emergence of the Edinburgh-based Court of Justiciary and its new ways of framing criminal charges. Scottish coroners remained active and important in enforcing law and order, and they did not wholly lose their functions until the reorganisation of central criminal justice in the late seventeenth and early eighteenth centuries. They were never formally abolished. Ultimately, the frictions between different frameworks of justice gave way, in the century after 1650, to a single set of Scottish officials, courts and procedures – in which coroners had no part.*

Houston, R. A. *The Coroners of Northern Britain c. 1300–1700.* Basingstoke: Palgrave Macmillan, 2014.
DOI: 10.1057/9781137381071.0009.

Some authorities have speculated that the office of coroner was obsolete by the sixteenth century.[1] Caithness is sometimes used as an example. The crown only acquired the earldom after the resignation (or possibly sale) by Alexander of Ard in 1375.[2] Kings held it until 1455 and, during that period, coroners became important as their representatives in the county. The office mattered enough to its holders for one, George Gunn, to sport the title 'The Crowner' (his daughter Christian styled herself 'Crownaris dochter'). The title invested its holder with authority, while physical symbol such as a brooch, wand, porteous roll or halberd (the coroner had the right to bear arms in and around the ayre) publicly intimated status. The justiciar himself had a viceregal role, but when the king went on ayre in person, as he did occasionally until 1602, the crowner's personal service was manifest. The coroner and the justiciar were the face of royal justice across late medieval Scotland, showing what the king's common law had to offer to barons and people alike. Even franchisal coroners were doing the king's business (and the right to appoint them came from the crown), making them important symbols and instruments of royal ambitions.

The appointment of justiciars and sheriffs also evidenced the significance of the personal in Scottish government and society, as the king invested judicial power in an individual justice, rather than the judge gaining authority from presiding in a court; Scottish justiciars had their own power bases in the areas of appointment (Scotia 'north of the Forth', Lothian and Galloway until the end of David II's reign) where English assize judges might have no such association and travelled on a greater number of circuits.[3] The crowner of Caithness is a signal example of the Scottish pattern. Occupant of the imposing Halberry Castle, George was also the seventh chief of the clan Gunn and his title may have been a blend of royal office and the kin-based authority of the *toíseach*. In 1455, however, William, the third Sinclair Earl of Orkney, regained the earldom of Caithness and when his son and successor later received the titles of hereditary justiciar and sheriff of Caithness a terrible feud began which ultimately led to the crownership falling into desuetude after the murder of the last holder, along with many of his kinsmen, at the Chapel of St Tayre, near Wick, in 1478.[4]

Yet the coroner is still in the documents during the sixteenth and seventeenth centuries. The printed Acts of the Parliaments of Scotland and the Registers of the Great Seal, Privy Seal, Treasurers' Accounts, Exchequer Rolls and Privy Council have been examined alongside manuscript court

DOI: 10.1057/9781137381071.0009

records and correspondence to find out where coroners can be found and what they did. As noted earlier, transfers of the office of 'coronator' become noticeable from the time of David II; in contrast, there are a few mentions after the Restoration and the last in these printed sources is the forfeiture by the Duke of Argyll of the coronership of 'Argyll and Tarbat' (1685). What evidence there is suggests continuing activity as judicial officers, with any significant decline not apparent until the late sixteenth or early seventeenth century.

There are indeed many examples of coroners at work (or expected so to be) and the office was significant enough for the crown to curb negligent or dishonest holders. In 1490 the King's Council summoned Fife's coroners to Edinburgh for instructions; the coroner of Forfar received a written reprimand in 1516 for not doing his job and the justice clerk of Dundee got an order to arrest 'certane gentlemen in fault of the principale coronour of Forfar'.[5] In February 1552 letters charged Peter Maccolm, one of the coroners of Wigtown, 'that misusit his office'.[6] Some coroners lost their position for malfeasance or insufficiency (variously construed), others because they were traitors or there was something wrong with their title, the latter a testament to the high social status of the office and its often proprietary nature.[7] Hereditary coroners found negligent could only lose their office and profits for a year and a day, while those not holding 'by fee and heritage' forfeited the office itself.[8] Any medieval or early modern office might be undermined by corruption or incompetence: this was an accusation often levelled at English coroners. Some objects of complaint may simply have lost the confidence of politically significant figures in their county; others used the job as a stepping stone to something better and may have cut corners.[9]

An assessment of a coroner's ability to implement the king's justice must, however, take into account three facts: conventions of compensation that could mean dropping a case once settled; the prevalence of arbitration, mediation and conciliation; the practice of 'repledging', where a suitably franchised lord could remove to his own court the case of an offence allegedly committed elsewhere by a person who normally lived within his jurisdiction. Compensation had always been a concern of Scottish justice, not only with felony as an offence against the crown, but also with *cro* or 'assythement' (compensation to kin), which continued to be important into the eighteenth century.[10] Advocating cases could also make a coroner's task thankless. The archbishop of St Andrews issued one of many commissions to his chamberlain and

DOI: 10.1057/9781137381071.0009

procurators to repledge his men from the justice-ayres in the spring of 1512.[11] Complaints about promiscuous granting of remissions too are frequent from the late fifteenth to the late seventeenth century.[12] In 1474, for example, the Scottish parliament deplored the 'gret derisione ande skorne of justice' manifested by offenders who chose to pay 'ane litill unlaw of silver' rather than undergo trial.[13] Exchequer records contain separate volumes of accounts for compositions (settlements of liabilities) from the justice-ayres of the 1570s and 1580s.[14] An Act of Sederunt (a set of procedural rules) of 13 October 1487 recognised further problems of enforcement, this time against offenders 'that he dar nocht nor is nocht of powere to arrest', ordering coroners to ask in the king's name for moral and material support from the offender's lord or a sheriff.[15]

In addition to removals for ineffectiveness, routine instructions suggest activity at a time of efforts to regularise and simplify the workings of justice. The references are admittedly sporadic and this may indicate that the functioning was irregular and intermittent – as was true of royal justice as a whole, which varied in its effectiveness over time and space.[16] For centuries, kings allowed territorial lords much greater latitude in law enforcement than was the case in post-Angevin England.[17] Yet until the seventeenth century the sense with coroners is of occasional mentions of routine functions, derelictions of otherwise competently conducted business or unusual problems in carrying out instructions. Both complaints and clampdowns came at times of political flux and are best seen as indicators of the importance attached to the proper administration of justice.[18]

That thin evidence obscures what was, rather than demonstrating what was not, is plain when we trawl the archives systematically. Detailed work by historian and archivist Athol Murray, on the earliest Justiciary Court books, shows a high level of expected activity in connection with justice-ayres south of the Forth during the reign of James IV.[19] Pertinent entries mainly concern fines on the coroner for failing to provide proof of either arrestment of those named on the porteous roll or of execution of summons or delivery. 'David Hume *coronator sepe vocatus ad intrandum Georgium Eskdale in Erssiltoun prout recepit eum in portuferio et non comparentem in amerciamento est primum iter in defectum probationis arreste eiusdem Georgii*.'[20] Sheriffs too might be fined for failing to enter persons named in the crowner's indenture and the close procedural supervision evident in this source suggests an active and important role for coroners in the late fifteenth and early sixteenth centuries. It is, of course, all too

DOI: 10.1057/9781137381071.0009

easy to point out the flaws in a system with uncertain central control: jurisdictions overlapped and sometimes competed, remuneration was based on results and exemptions, traditions of compensation and lordly influence all served to limit effectiveness. Hunnisett's work on English medieval coroners dwells on exactly the same alleged shortcomings.[21] When activated, Scotland's ayres and their officers nevertheless seem to have functioned effectively.

Justiciary records are important in taking analysis beyond the repetition of clauses from legislation, which form the backbone of most attempts to shed light on the coronership. References are less frequent in other sources, but the same sense of routine functioning with occasional failings emanates from the documents until the early seventeenth century. Justice rolls were sent to the coroners of Dumfries, Kirkcudbright, Wigtown, Fife and Angus (Forfarshire) in 1502, the coroner of Ayrshire in 1509, Dumbarton in 1539 (accompanied by a messenger at arms who stayed to oversee implementation), the coroners of Peebles and Selkirk in October 1573 and 1576, Aberdeenshire in 1574, Dumfries and Annandale in 1575.[22] The coroner of Dumbarton was important enough for the Exchequer and Lord Justice Clerk to spend 12 shillings on sending a boy with a packet of letters in January 1568; they dispatched another boy in March 1596 to the sheriff clerk of Dumfries with brieves to summon assizes (criminal juries), deliver dittays (indictments) and warn the 'coronell' to attend and receive the porteous roll from the clerk in preparation for a justice court to be set up there the following month.[23] As coroner of Lothian in the constabulary of Haddington, the laird of Edmonstone received frequent letters from the Lord Treasurer in the late 1560s and early 1570s, including a 'blude roll' in November 1573.[24]

What of the job's rewards? Coronerships, which were often hereditary, mattered to the holder, not only because of the status they conferred and confirmed (or even expanded beyond the territorial bounds that normally circumscribed it), but also because of fees and perquisites. Early coroners proper may not have been paid, but the fourteenth-century 'Laws of Malcolm MacKenneth' state that fees, amercements and escheats of justice (forfeitures of moveable assets) should go to the crowner *'ubi qui accusat adiciat in appellatione sua pacem domini Regis fore infrictam'* (where he who accuses should add in his appeal that the peace of our Lord king shall be broken); this may have been more for accounting than remuneration and, as the laws were written by someone who disliked private justice, they may have been as argumentative as

DOI: 10.1057/9781137381071.0009

Sir John Skene's comments.[25] In later times, fees are clearly documented and might be justified nationally through legislation or locally through custom. According to Balfour, a coroner was due 2/6d 'for ilk man unlawit and fylit' along with certain specified livestock, grain and household goods.[26]

Balfour's authority was the 'Laws of Malcolm MacKenneth' and most of the later documentation concerns local usages. The Lords of Council confirmed John Home of Cowdenknowes, coroner of Berwickshire, in his entitlement to unlaw due by the town of Lauder in 1529.[27] The following year the freeholders of the barony of Renfrew tried to argue that their coroner was only entitled to a quantity of grain per plough and a flat rate of 4d per household (as suggested in a judgement of 1494), but the Lords of Council found that he was also due half a merk for everyone who compounded or had been remitted at the last ayre.[28] In 1594–5, Duncan Forbes received £5 from the provost of Aberdeen's order 'for his crownar fee for the haill toune, quhen thai suld have enterit to the justice air'.[29] A Court of Session judgement in M'Lay or Mackay v. laird of Skelmorlie (10 July 1623) upheld the coroner of Arran's entitlement to regular payments from heritors, provided he could prove that these had been customarily made.[30] James Stewart of Torrence, serjeant, coroner and governor of the regality of Kilbride, 'conforme to old use and wont' got half a boll of corn for every plough, a firlot from everyone who sowed, but had no plough, half a merk 'or the uppermost garment' for every sasine, a cheese from every household, and finally 2/- from every amercement levied by the court.[31] How lucrative these emoluments were in total is unclear and litigation over them may say more about honour and power than money.[32] The Earl of Kinghorn was entitled to £4 sterling from the justice-ayre processes when confirmed in the office of crownary of Forfar and Kincardine in 1672 – a sum unchanged since 1382, which suggests the status of the office mattered more than its fees.[33]

The Earl of Kinghorn could be relaxed about collecting his £4 sterling – if not about his entitlement to it. Scottish titular coroners were nearly always members of the landed nobility, sometimes aristocrats. The office could be a mark of royal favour on its own or in combination with other awards. For example, Sir John Edmonstone of Edmonstone appears as a person of considerable importance when appointed by David II as coroner to the shire of Edinburgh in fee, accompanied by grants of lands in Banffshire.[34] Those who had the title of coroner in Scotland had to have political and social clout in their own right, as well as that borrowed

DOI: 10.1057/9781137381071.0009

from the monarch or lord who appointed them. Not always members of affinities, they still had a part to play in local power dynamics. In a mutually reinforcing relationship, holding the office added to a man's store of honour and power.

This could, however, be done in such a way as to bind him more effectively to the donor. The Forbes family held the deputy coronership of Aberdeenshire during the sixteenth and early seventeenth centuries as a way of cementing loyalty to their traditional rivals, the mighty Gordon Earls of Huntly, who were the hereditary sheriffs.[35] In Argyllshire, by contrast, the power of the Campbells was absolute and the Dukes there could appoint their own cadets to perform the role. That an acting Scottish coroner could be an appointee of a subordinate feudal lord as well as of the monarch is important, for those with deputed rights of public justice retained the privilege of delegating powers that had been lost by most English barons in the twelfth century.[36] Early justiciars too could appoint deputies as well as the king.[37] Scottish kings thought they had more to gain by allowing a man of proven power and influence to recruit weighty adherents to the crown's service than they lost to self-interest by allowing delegation.[38] For example, the Earl of Fife made Duncan, Earl of Lennox, coroner of Lennox in the early 1390s.[39] As late as 1562 John, archbishop of St Andrews, appointed Patrick Learmonth of Dairsie to the offices of bailie, steward and justiciar, and coroner of the courts of the justice-ayre of St Andrews regality: a long list that must itself have required delegation.[40]

Disputes about title confirm the importance incumbents attached to being coroner. In 1556 David, Lord Drummond and John Creichtoun of Strathurd, who disputed tenure of the office, threatened John Tosheoch of Cultre and others in the stewarty of Strathearn with poinding for coroners' fees of half a merk each on their lands.[41] In 1582 Lord Innermeith, hereditary coroner of Perthshire 'with the proffites and feis of the same', had complained to the Privy Council about deforcement (obstruction) of his officers at Dunblane burgh cross by Lord Drummond's men while executing summons to appear at the justice court of Perth. In fact, the justice clerk had delivered the porteous rolls to Drummond's late father and the Privy Council allowed him to continue as coroner until the Court of Session determined their case.[42] The Privy Council heard another dispute, this time in 1597, between the lairds of Machrimoir and Spotts over who was the hereditary coroner between Dee and Cree, the decision made locally at Dumfries during a judicial expedition.[43]

DOI: 10.1057/9781137381071.0009

These contests may simply reflect chronic tensions between houses, but they may also signal their changing fortunes. Families whose wealth and influence faded might surrender the coronership, suggesting that the office itself could not compensate for a lack of personal status and authority. For example, McDowell of Machrimoir had to pay 20 merks composition for resigning his 'crounareschipe and hed of kyne [*toíseach-deor?*] in the partis of Gal[o]waye' in 1473–4 and the Duddingstons of Southhouse, coroners of Peeblesshire for much of the sixteenth century, lost the office to the Murrays during the 1570s.[44] Robert Graham of Knockdolean, crowner of the whole sheriffdom of Dumbarton and of that part of the sheriffdom annexed to Stirling, entered into a contract to sell it to Sir John Colquhoun of Luss in 1569.[45]

There were other reasons for surrendering a coronership. Presenting a crime was easy enough compared with arresting a criminal and securing his goods. Justice-ayres dealt with some of the most difficult cases in the kingdom and in consequence coroners' jobs could be manifestly dangerous, not just in military matters or the risky business of curbing vagrants. Coroners had to know where the accused lived and take surety from him; they had to know who had seizeable assets in their jurisdiction if the accused could not be located in person and, as they had to imprison those without goods to be arrested, they also had to be abreast of how dangerous their quarry could be.[46] A draft proclamation for the justice-ayre of Jedburgh in 1511 lists 'schireffis, crounaris and thair deputis' among those licensed to bear arms and the justiciar, justice clerk and others could be required personally to attend judicial raids and musters in the sixteenth-century Borders.[47]

Even with prior knowledge and assistance, the work could be unpredictable. The Master of Montgomery wounded William Cunningham while discharging his office in 1505 (Montgomery was summoned for treason for the offence) and the following year the crown licensed the Earl of Argyll and Lord Montgomery to associate and support George Railstoune and James Striueling, who were 'at the horn' (outlawed) for the assault.[48] Argyll was a coroner in his own right, reinforcing the impression that the reason for the attack was status and jurisdiction, for there had been major confrontations between the two families in and around Irvine during the 1480s (when William Cunningham of Craignish became coroner of Wester Strath Gryfe and of Over Ward of the barony of Renfrew) and 1490s; the Cunninghams of Caprintoun surrendered the coronership of Cunningham to Hugh, Earl of Eglinton, in

DOI: 10.1057/9781137381071.0009

1535, but kept those of Wester Strath Gryfe and Over Renfrew.[49] Gilbert Graham of Knockdoliane, principal coroner of Dumbarton, was robbed, held to ransom, and his porteous rolls torn and burned while discharging his duties in the lawless area of 'Drummond' (parish of Drymen, Stirlingshire) in 1532.[50]

Complainants may have played up deforcement of royal officers in instances where something else was at stake. Yet these examples make it clear that the coronership was no mere sinecure, entailing time-consuming, responsible, difficult and sometimes hazardous tasks. That holders were supposed to be active is further evident when the job was bifurcated. A distinction between a titular and acting coroner is particularly obvious when women held the post as heiresses or in conjunction with their husbands. During the fifteenth century Agnes Vaus was coroner of half of the barony of Renfrew and Alison Park coroner (and mair of fee) of the ward of Strath Gryfe.[51] In a Privy Seal grant of 1529, we find that Mr Duncan Makke (Mackie) of the Larg and his wife Janet Gordon had the wardship of the lands and profits of the late Uthred McDowell of Machrimoir, including the office of 'crownarschip' between Dee and Cree. Duncan died and Janet remarried David Crawfurd. Thus, James V deemed that David should hold the office during the wardship, 'the said Jonet quhilk is ane woman nocht able to exerce the sammyn'.[52] Other orders and charters similarly suggest that the coroner was one person, he who performed the office someone else.[53] Men (and women) might derive honour from a coronership (and lend the weight of their status to the office), but they had to ensure that the job was done to the king's satisfaction.

Much earlier, an act of 1357, which required persons infeft in such offices, but who were not doing their duties, to present suitable substitutes to the king, also suggests a division between honorific role and functional duties.[54] In 1430, mairs of fee were empowered to appoint deputies, who may have been called mairs, serjeants or officers.[55] The Earl of Mar and Garioch granted a charter including the heritable coronership of Garioch to John Blackhall of Blackhall in 1433. However, an earlier decreet absolvitor (a judgement in favour of the defender) from the regality court of Garioch mentions him as the holder in 1418; James II confirmed the post in 1457, a further reminder that verbal appointments might long precede patents and that written records are not a reliable guide to officeholders on the ground during the Middle Ages.[56] In 1536, the king gave Oliver Sinclair *gratis* lands and made him coroner of

DOI: 10.1057/9781137381071.0009

the sheriffdom of Peebles, following the death of William Duddingston of Southhouse and a non-entry of 23 years.[57] In 1572 Archibald, Earl of Argyll, granted Colin Campbell of Barbreck land and the coronership of Glenorchy.[58] A letter to George Home of Aytoun in 1573 describes him as 'crownar principall of Berwik' and William Cranston of Cranston in Roxburghshire had the same title, indicating there were also subordinates.[59] John Grahame subscribed a Perthshire justice roll of 1582 as 'crowner depute' and James Hay a porteous roll of 1628 as 'crownar deput' for the burgh of Perth.[60] The mid-seventeenth-century coroner of Bute must have had a deputy as his main landholding was in Ireland.[61] An order of 1600 to the coroners of Perth, Strathearn, Menteith, Forfar and Fife, to answer charges of corruption, mentioned an elaborate group of assistants also required to present themselves before the Privy Council in Edinburgh. These included 'thair deputis officers and clerkis quha wer imployit be thame in the matter of the iustice court'.[62]

Aristocrats from the monarch downwards liked having titles, but they did not always want to soil their hands with the day-to-day performance of the tasks they entailed. Passing the duties on to someone else did not, however, absolve lords from responsibility and principal coroners were financially liable for the actions of their deputies. In 1503 Alexander, Earl of Buchan, pursued his deputy, Alexander Stewart of Kinmachlon, for failing to indemnify him against surrendering to Exchequer unlaws and a composition for remission at the recent justice-ayre.[63] The Earls of Buchan seem to have been chronically unlucky with their choice of deputies. On 13 April 1548 John, Earl of Buchan, hereditary sheriff and crowner of Banff, appointed Sir Walter Ogilvy of Boyne as his deputy for a term of five years. Ogilvy agreed to warrant the Earl at all hands 'of all maner of charge' (in other words, he became accountable to Exchequer). Either Ogilvy pocketed the proceeds or did not make seizures, for in January 1549 Buchan instituted proceedings against him for failing to poind for the unlaws of the justice-ayres, for withholding court books and for usurping the function of principal (he wanted Ogilvy to cease acting as deputy coroner). In 1552, Ogilvy was put to the horn for failing to produce his court books or to make account and again ordered to produce them in July 1553. Buchan accused him of not paying his taxes in August of that year, but Ogilvy produced a receipt and this seems to have been a vexatious accusation.[64]

The office was certainly honourable, but it was not merely honorific, entailing responsibilities as well as rights. The crown disciplined holders

DOI: 10.1057/9781137381071.0009

and closely regulated terms of office. Directing coroners away from an area tells as much about their importance as does allocating them to one. David II excluded royal coroners from the regality of the abbey of Arbroath and from interfering in the bishop of Moray's right to try crimes committed by his tenants in Strathspey and Badenoch.[65] As late as 1553 the bishop was able to prove that his tenants in the regality of Spynie had never paid a crowner's fee to the sheriff of Morayshire, but to the regality's crowner; grants of regality entitled the holder to appoint officers parallel to those of the crown.[66] Because of what they did and how much they charged both criminals and blameless inhabitants liable for contributions in kind, coroners were no more popular with the laity too than other court officers like messengers or heralds. An act of parliament (1449, c. 8) stopped them from taking 'wrang custum' or payment from accused people who found adequate surety, and an act of 1487 (c. 7) regulated how much they would receive from sheriffs as a share of criminal forfeitures.[67] Parliament put procedures in place at the same time to try any complaints against a coroner on the last day of the justice-ayre and to punish him if convicted.[68] This may be why a late-fifteenth-century poem condemned the extortions of 'Justice, Crounar, Sariand, and Justice Clark'.[69] Criticisms of coroners were, nevertheless, usually *ad hominem* and the office itself was not the subject of general condemnation or ridicule until the seventeenth century. Fifteenth-century legislation aimed to prevent abuses and promote confidence rather than indicting the system as a whole.[70] Like much contemporary legislation more generally, it may have been less a deliberate attempt at reform and more a declaration or sanction of what actual practice had proved expedient.

Most of the examples given earlier come from the comparatively well-documented fifteenth and sixteenth centuries, which give the impression, possibly misleading, that this was the office's (and perhaps the ayres') heyday. In contrast, declining mentions in the Register of Deeds and Exchequer Rolls from the 1580s may be an indicator of its decline.[71] Whereas political and military roles became more prominent in the sixteenth century, other functions waned. In his early-seventeenth-century *Major Practicks*, Hope could still describe the coroner as 'the justice officer, and makes arreistments or attachments upon all persons that ar to be indytted and accused befoir the justice in justice airs … the catalogue of the persones names that ar indytted wes delivered be the justice clerk to the crouner, that he might arreist and attache them conforme thereto and is called ane porteous [roll]'.[72] Historian Michael Wasser finds, however, that indictments, initiated by a private complaint to the crowner who organised summons to the justice-ayre, were unusual in the early-seventeenth-century Edinburgh-based justice court. Instead, it required criminal letters addressed to the justice clerk and handled by a messenger at arms, of whom there were supposed to be 183 'apportioned among the different sheriffdoms'.[73] These were probably much more numerous than coroners (though possibly not more than their deputies); Edinburgh alone – admittedly the legal hub of Scotland – had 22 messengers resident in the city during the 1690s.[74]

Once known as 'officers at arms' or simply as the king's officers, messengers probably originated during the twelfth century and were firmly established in the late fourteenth century. Their work becomes visible only in the time of James IV, notably when deforced.[75] Formally regulated in the 1520s, 1540s and again in the 1570s, legislation of 1587 constituted them as executors of all the king's letters in both civil and criminal causes. They gained clear responsibilities, oversight and a *per diem* payment, changed to a levy on pursuers in the seventeenth century.[76] The century after 1587 saw procedure and accountability further tightened in ways similar to the fifteenth-century legislation on coroners, albeit in a very different political and legal climate. An early-nineteenth-century treatise on the office of messenger, by Alexander Frazer, clearly shows the overlap between these officers and historic coroners. 'The duty of a messenger consists, chiefly... In writing and executing copies of citation upon summonses, charges upon hornings, &c. arrestments, inhibitions, and other diligences, and extending proper executions thereon... In executing poindings... In apprehending, and incarcerating, persons upon captions, and other legal warrants'.[77] Having said this, late-medieval messengers had a distinctive remit in addition to their legal work, listed by historian Katie Stevenson as 'carrying... diplomatic briefs; organising and participating in royal ceremonies, such as coronations, weddings, funerals and tournaments; acting in matters as advocate for the king and as procurators for private clients; and collecting the heraldic information of the kingdom'.[78] Coroners sometimes needed the help of local sheriffs and landowners in their work – messengers too, including one deforced during the 1550s while poinding the goods of some Aberdeen burgesses for sums due to the coroner as a result of a justice-ayre.[79] For their part, messengers carried the king's letters to the greatest in the land; during the sixteenth century they began to proclaim at the mercat crosses of

DOI: 10.1057/9781137381071.0009

royal burghs, in time becoming the normal means to make royal procla-mations.[80] When, in June 1600, the Privy Council summoned the coro-ners of Perth, Strathearn, Menteith, Forfar and Fife personally to appear before it to answer charges of 'extortioun and skaffae [skaith? (damage) or perhaps scavenging or sorning?]' in connection with the ayres held that spring, Robert Elder, messenger, delivered the letters.[81]

Differences in their symbols of office further distinguish the men. The coroner had a porteous roll, halberd or other weapon(s), wand, possibly an unspecified brooch and, although there is no explicit record of one, a horn to summon the shire to his aid. None of these was a legal requirement. They were the symbols of the king's justice, notably the porteous roll for criminal indictments, which were specific to the ayres and were not used by franchise courts.[82] Some of these objects were, on the other hand, also instruments of force for men of action. In contrast, the messenger sported a badge (made of silver from 1587, with the king's arms upon it – thus his title), a small but elaborate rod, baton, or 'wand of peace' (at six inches, this was much shorter than the two feet three inches specified in 1432), his letters, a stamp or seal and a horn to denounce 'rebels' or outlaws and make public his actions; for important proclamations at market crosses he might be accompanied by a herald and trumpeters. The only other requirement was a horse because messengers had to (and could) travel anywhere in Scotland whereas coroners (and sheriffs and sheriffs' officers alike) had to stick to their nominated or 'resident' jurisdiction. Legislation specified each of the messengers' items.[83] They were his defences, in the same way as the accoutrements borne by the coroner, but they were the symbols of authority, justice and peace.

When jurist Sir George Mackenzie of Rosehaugh wrote his observation on statute 99 of 14 James III, the judicial and administrative environment truly had changed. He felt sure that 'the Crowner of old received the Porteous Rolls, that is to say, the names of such Malefactors as were to be pursu'd at Justice-airs, but now the Justice Clerk keeps it himself, and gives it to the Macers of the Criminal Courts, or Messengers who cite the persons to be pursu'd'. Of act 102 he added that coroners no longer attached and thus should not get fees, which explains why he further noted in connection with statute 5 of 3 James V that current holders complained about withholding the porteous rolls from them.[84] Rather than a simple story of centralisation, the fate of coroners exemplifies how political change in early modern Scotland involved one type of centralised judiciary (the Edinburgh-focused Justiciary Court and its officers) taking over from another (the justice-ayre and the coroner).

The reconstitution and reorganisation of courts in the 1520s and 1530s may explain why Pinkerton thought 'the *crowner*, or coroner, continues to attract notice even in the reign of James V, with which his ancient office seems to expire'.[85] Coroners' role as part of justice-ayres had been diminishing since the early years of the sixteenth century along with that of the ayres themselves, held increasingly to raise money (especially through the sale of remissions) or to make specific political or military points; it is possible that the ayres had always been sporadic.[86] A central 'justice court' took over more of their regular work after 1524–5, later called the Justiciary Court, with an enhanced role for the justice clerk. Litigants drove some of this development and commentators recognised that 'the proces of justice aires is sa long and prolixt that in mony zeires parties that ar hurt and grieved gettis na justice'; changes towards peremptory procedure further limited coroners' roles.

There was, however, no simple displacement of the ayres by the Justiciary Court for much of the sixteenth century. Attempts to revive and extend justiciars by legislation of 1587 c. 57 failed and they held ayres only intermittently thereafter.[87] Orders came for justice-ayres on designated days in some parts of Scotland in the late 1590s and early 1600s, including the sheriffdom of Berwick and Lauderdale, Dumfries, Fife (at Cupar), Forfar (at Dundee), Lanark, constabulary of Haddington, stewartries of Mentieth and Strathearn, Peebles, Perth and east Teviotdale.[88] Yet even in these cases, the business of turn-of-the-century ayres became more restricted and special commissions of justiciary increasingly took business away from them. In 1628, when the crown attempted to revive ayres after the resignation of the last hereditary Justice General (Archibald, Earl of Argyll), the Privy Council took steps to remedy the problem of shires without coroners, suggesting some had disappeared from certain areas, along with the ayres they served.[89] Contemporaries sensed change. A group of Englishmen who visited southern Scotland in 1629 noted: 'The last year, 1628, the Judges went circuits, but it is doubted whether they will hereafter do so or not.'[90]

The Interregnum marked a decisive watershed in introducing a more streamlined version of criminal circuit courts, pruned of superfluous offices. Abolished by Oliver Cromwell in 1652, Charles II reinstated heritable coronerships, but they never recovered the exercise of their powers.[91] Mackenzie thought 'this Office is absolet [obsolete] now, except

DOI: 10.1057/9781137381071.0009

at Justice Airs', its functions having passed elsewhere: for example, the Justiciary Court or the Sheriff Court usually issued criminal writs directly.[92] Mackenzie wrote during the eclipse of the office thanks partly to the focusing of royal criminal justice on sheriffs and the Justiciary Court and partly to the rise of justice clerks, messengers and procurators fiscal. Crowners were, nevertheless, still active in searching out and seizing stolen or forfeited goods and serving summons in Argyllshire during the 1660s and 1670s.[93] The same source, Argyll Justiciary, makes it plain that procurators fiscal and coroners were not the same individuals and that they performed quite different tasks.[94]

Mackenzie was unsure about procedures outside his direct knowledge, yet there was still life in the office of coroner in his day. In 1670 the Earl of Argyll appointed John Campbell, younger, of Glenorchy (in 1681 the first Earl of Breadalbane), as coroner-depute within the lands and barony of Glenorchy and lands of Cataneis and Larig (otherwise called McLachlan's lands). This was an appointment, about the conduct of which we know nothing. In contrast, a hotly contested legal battle fought during the decade after 1679 sheds considerable light on the nature and fortunes of the late-seventeenth-century coroner. At its heart was a dispute over fees and the ownership of the office between the sheriff of Bute (partly as sheriff, partly as landowner or 'feuar' there) and Mr John Steuart of Ascog, advocate, who held the office of hereditary coroner there. The issue centred on payment of a lamb and firlot of corn by every tenant on feued land, a due confirmed in a judgement by the Commissary of the Isles in 1667.[95] In 1679 the sheriff issued an order 'upon a Sabbath day betwixt sermons by his officer', banning payments to Steuart as coroner of Bute. The sheriff also allegedly attempted to get the advocate to sell him the office, which had until the 1660s been tied to lands that the sheriff had since acquired.[96] In 1655 Robert Kerr agreed a price with Robert Jamesone for his lands and the office of 'crounarie'; Jamesone disponed (disposed of) the lands to John Boyl of Kelburn who in turn sold them to James Stewart, sheriff of Bute. Ker disponed the office separately to Steuart of Ascog in November 1666, a few months after Steuart's admission to the faculty of advocates.[97]

For his part the sheriff alleged (among much else) that: statute allowed for the coroner's remuneration 'not out of the goods of Innocent people … but out of the goods of the Guilty'; 'the Casualties as they are now craved are most unreasonable in themselves for the exaction now pretendit to is thus: If on[e] rich tennent hes a pleugh [plough] he payes

DOI: 10.1057/9781137381071.0009

but on[e] lamb & on[e] firlot of oats & if ther be ten poor tennents about ane pleugh each on[e] of them payes as much which is most inconsistent with equitie & reasone and showes that the said exaction is foundit upon noe law'; regardless of how long the casualties had accrued, it was unreasonable for Ascog to expect them when 'the Crouners office is absolete & in dissuetude [disuse] and the sheriff himself serves for him by attaching and presenting prisoners which was all that he was bound to'.[98]

One might note, in passing, that Ascog was unusual since Scottish coroners were only incidentally lawyers (he got the office conveyed to him in the 1660s in exchange for representing the then owner in legal contests) whereas a legal background was common among early modern English coroners.[99] Naturally, someone implementing the king's common law would become well-acquainted with judicial proceeding and the practical working of the law, but (like medieval Scottish justiciars) coroners' lack of formal training remains clear. Again, this carried on the medieval practice where men of social importance rather than legal knowledge were the senior officers in the Scottish king's courts; native legal experts were a different class of men known as *judices* in Latin documents. Only in the late sixteenth century, with the formal regulation of messengers, did officers have to be men with 'sufficient knowledge, learning, and experience, for executing the office' of serving the king's letters, though it is likely that many had received a formal education in the law even in the late Middle Ages; the foundation of Aberdeen University in 1495 and an education act of 1496 came out of the push for better educated élites to fill such posts.[100]

In England, the crown found 'a corps of itinerant justices' from the ranks of professional lawyers no later than the thirteenth century.[101] One of the reasons English Justices of the Peace replaced keepers of the peace during the fourteenth century was because of the latter's lack of legal training and, by Elizabethan times, Sir Thomas Smyth could assert that the coroner too was usually 'a man seene [versed] in the lawes of the Realme'.[102] To take another example of the different chronologies of 'legalisation', officers at arms were professionalised in England fully a century before their Scottish counterparts.[103] With Inns of Court to train them, lawyers became ubiquitous in many branches of English economic and social affairs as well as administrative life. In time formal legal training became essential to Scottish messengers and Robert Thomson's 1753 *Treatise* could specify that the holder should be 'a Person of Discretion, Honesty and Credit', whereas of the office he opined: 'besides a reasonable

DOI: 10.1057/9781137381071.0009

stock of Prudence and Experience, it requires considerable Knowledge in Law, and the Art of forming Writs'.[104]

Whatever the perceived archaism of the coronership in Scotland, there were still appointments and expectations in the late seventeenth and early eighteenth centuries, notably in the regalities of the west.[105] Alexander Fullerton of Kilmichael became coroner of part of Arran in 1684 'with provision that the customary services of the coronership are to continue to be performed' and his son James succeeded him in 1711.[106] In the Highlands as late as 1709, armed men mobbed Archibald Campbell of Barbreck, hereditary coroner and bailie of mid-Lochawe (Argyllshire), as he was trying to find and attach them.[107] Even outside these jurisdictions and areas there were continuing signs of activity. In the south-east Alexander Home of Aytoun, hereditary coroner of Berwickshire, issued a commission to inventory the goods of a suspected Eyemouth witch in 1662.[108] In the south-west the king still thought it worth removing Patrick Dunbar of Machrimoir from the coronership between Dee and Nith in 1681, this during one of the post-Restoration periods (1679–84) with documented circuit courts or ayres.[109]

The trend towards obsolescence is nevertheless plain and an act of the Westminster parliament in 1709 meant 'abolishing the method of exhibiting criminal information by the porteous roll'.[110] Prior to this, prosecutions had been based on indictments. Central government sent out formal letters to sheriffs, asking for a list of suspects of specified crimes. An inquest questioned local worthies and their answers went into the porteous roll. After 1709, various jurisdictions and officers 'delated' (reported or charged) cases, instead of using dittays.[111] This may have marked the end of any practical significance for the office of coroner and it is not mentioned in the Heritable Jurisdictions Act implemented in 1748, suggesting that it was not thought worthy of compensation in its own right.[112] The Court of Session dismissed a claim for recompense of £500 sterling, made in 1747 by the hereditary coroner of Forfar and Kincardine, on the not unreasonable grounds that he should not be compensated for something that had not been taken away from him.[113] When jurist David Hume wrote his *Commentaries* in 1797, he described coroners as 'now unknown to our practice'.[114]

Yet the office continued to exist as a hereditary position in some subsequent charters and other documents. For example, in April 1798 Dugald Campbell of Craignish issued a disposition to trustees for his creditors of the lands of Lagganlochan and Ardlarach (Argyllshire),

DOI: 10.1057/9781137381071.0009

with the office of bailie and coroner of said lands.[115] The fees, if not the conduct, of the office were still allegedly in place 'conform to ancient use and wont' when in 1838 the *curator bonis* (court-appointed administrator; like an English guardian or committee of the estate) for Miss Christian Anne Stuart of Torrence pursued dozens of inhabitants there for non-payment of the profits of the heritable office of serjeandry and coroner of the lordship and regality of Kilbride (a given measure of oats for each plough) unpaid since 1806.[116] The sense is of a traditional office gradually sidelined by new ways of organising justice: an anomaly whose incumbents continued to try to gather in fees – even if their functions had been supplanted and their links to central government weakened by disuse.[117] As jurist James Dalrymple, first Viscount of Stair, observed of the laws of Scotland in general, 'what is found inconvenient is obliterated and forgot.'[118] Coroners had not always been failures waiting to be swept away in favour of something better, for they had been manifestly active and useful, especially in the fifteenth and sixteenth centuries. Instead, the crown used different means to achieve extensive central control of the mechanisms of justice and the society which the criminal courts served and controlled. Lawyers too reconfigured criminal justice to focus on the justice clerk and messenger, and later the Justice of Peace and constable. The title of coroner alone remained for those whose charters had correctly conferred it.[119] In the 1800s advocate Gilbert Hutcheson could observe that 'though the name of hereditary coroner still adorn the titles of some ancient families, yet the office having long ago lost its power and jurisdiction, gives no charge of the peace', concluding that it was now 'mere *vox et pr[a]eterea nihil*' (nothing but a term).[120] In the present day, the title itself is obsolete.

Notes

1   W. C. Dickinson, 'The toschederach', *Juridical Review* 53 (1941), 85–109. W. D. H. Sellar, 'Celtic law and Scots law: survival and integration', *Scottish Studies* 29 (1989), 8–11. W. Gillies, 'Some thought on the Toschederache', *Scottish Gaelic Studies* 17 (1996), 128–42.

2   B. E. Crawford, *Northern earldoms: Orkney and Caithness from AD 870 to 1470* (Edinburgh, 2013), 235–6.

3   Wasser, 'Central criminal courts', 94, 99. *Ancient Scottish poems*, 113. NAS JC1/2, f. 230v. Walker, *Legal history of Scotland* vol. 1, 221. Davies, 'Scottish

DOI: 10.1057/9781137381071.0009

legal system', 134. That the justiciarship was not heritable is clear no later than the mid-fourteenth century, when there was considerable turnover of personnel. MacQueen, *Common law*, 62–3.

4   Gunn, *Clan Gunn*, 43–57. M. J. Gunn, 'The clan Gunn and the office of crowner in Caithness', *Clan Gunn Society Magazine* 16 (1980–1), 33–9. G. Neilson and H. Paton (eds), *Acts of the lords of council in civil causes* vol. 2 (Edinburgh: HMSO, 1918), 481. A. MacGregor, *The feuds of the clans* (Stirling: E. Mackay, 1907), 69–70. The office still appears in subsequent charters. NAS GD112/58/1/2 (1526).

5   ATS I, 134. Arrests had to be witnessed across all of Scotland after 1474 (c. 7). APS II, 107.

6   ATS X, 62.

7   Edward I's understanding of satisfactory or 'suffisantz' coroners was those who had proper title or 'chartres du doun des offices'. Stones (ed.), *Anglo-Scottish relations*, 123–4.

8   K. M. Brown, A. J. Mann and R. J. Tanner (eds), *The Records of the Parliaments of Scotland to 1707* (St Andrews: University of St Andrews, 2007–2009), 1458/3/24.

9   Hunnisett, *Medieval coroner*, 118–26. Cam, 'Shire officials', 150–2, 159. Lloyd, 'Coroners of Leicestershire', 21–3, 29.

10 Grant, 'Franchises north of the border', 179. NAS JC3/1, James Somervell (December 1704). *Arguments and decisions in remarkable cases before the High Court of Justiciary and other supreme courts in Scotland. Collected by Mr MacLaurin* (Edinburgh: J. Bell, 1774), 99. Hume, *Commentaries*, vol. 1, 44, 286. K. M. Norrie, 'The intentional delicts', in K. Reid and R. Zimmermann (eds), *A history of private law in Scotland. Volume 1: introduction and property* (Oxford: Oxford UP, 2000), 484–8.

11 NAS GD45/13/303.

12 G. H. W. Gane, 'The effect of a pardon in Scots law', *Juridical Review* (1980), 19–20. MacQueen, *Common law*, 61. Nicholson, *Scotland*, 430–1. A remission reduced or removed a sentence, usually in exchange for a 'composition' or cash payment, while leaving the conviction to stand.

13 APS II, 107.

14 NAS E22/15 and 16.

15 Brown *et al.* (eds), *Parliaments of Scotland*, 1487/10/8. APS II, 176, 177.

16 Wasser, 'Central criminal courts', 96.

17 Grant, 'Franchises north of the border', 155–99.

18 MacQueen, *Common law*, 54–5, 59–60.

19 NAS JC1/1–2. I am most grateful to Dr Athol Murray for sharing this material with me.

20 NAS JC1/1, p. 9 (9 November 1493). The coroner had often been called upon to produce George Eskdale, cited in the porteous roll, and was fined for failing to prove he had done so.

DOI: 10.1057/9781137381071.0009

21 Smith, 'Medieval coroners' rolls', *passim*, offers a more positive view of
the work of coroners and a more optimistic assessment of the patchy
documentation they have left.

22 ATS II, 133, 136; III, 135; VII, 261; XII, 364; XIII, 25, 85, 141, 147. Messengers are
discussed more fully later.

23 ATS XII, 97. NAS E21/70, ff. 196r, 197r. The precepts of justiciary and
summons to the steward of Kirkcudbright were separate from those to
the sheriff clerk of Dumfries, and the courts for Dumfries, Wigton and
Kirkcudbright were held on three different days. MacQueen, *Common law*,
63–4, describes procedure, but makes no mention of coroners. In February
1598, precepts were issued to coroners of Roxburgh, Selkirk and Peebles to
summon those mentioned in the porteous roll to a justice court at Peebles.
NAS E21/72, f. 48v.

24 ATS XII, 15, 37, 48, 72, 145, 156, 172, 204, 222,229, 235, 277, 284, 305, 329,
333, 370, 384. The blood roll went to the sheriff of Perth in 1566, a separate
coroner's roll to the crowner of Perth. ATS XI, 465.

25 APS I, 737. R. L. G. Ritchie, *The Normans in Scotland* (Edinburgh: Edinburgh
UP, 1954), 321. Duncan, 'Laws of Malcolm MacKenneth', 244, 252, 256–7.
Grant, 'Franchises north of the border', 174.

26 *Balfour's Practicks*, 566–7.

27 Hannay (ed.), *Lords of council, 1501-54*, 313.

28 *Acts of the lords auditors*, 192. R. K. Hannay (ed.), *Acts of the Lords of Council
in Public Affairs, 1501-54* (Edinburgh: HMGRO, 1932), 329. *Archaeological and
historical collections relating to the county of Renfrew, parish of Lochwinnoch* 2
vols (Paisley: A. Gardner, 1885–90), vol. 1, 146–7.

29 'Extracts from the accounts of the burgh of Aberdeen', *The Miscellany of the
Spalding Club, volume fifth* (Aberdeen, 1852), 118.

30 Morison, *Dictionary*, 10,887. Martin, *Western islands*, 224. A heritor was
strictly a landed proprietor liable to contribute to the upkeep of the parish
church, though in practice the word could be applied to any landowner.
On the meanings of 'custom' in Scotland see R. A. Houston, 'Custom in
context: medieval and early modern Scotland and England', *Past & Present*
211 (2011), 35–76.

31 APS VII, 609. Other seventeenth-century documents describe the office
as both a 'coronership' and a 'mairship and colonellship and heretable
governement', separating the sources of the fees according to the specific
office. NLS Ch. 7981 (1652). See also NAS C16/1, ff. 40v–41v. Eighteenth-
century sasines and charters describe the holder of 'the heretable office of
Serjandry & Crownership & heretable Government'. NLS Ch. 8008 (1736).
Ch. 8017 (1774).

32 For another example of a power struggle that used fees payable to a coroner
as an issue see *Historical notices of Scottish affairs, selected from the manuscripts*

of Sir John Lauder of Fountainhall, bart. 2 vols consecutively paginated (Edinburgh: Bannatyne Club, 1848), 624, 847.

33 APS VIII, 193b. However, the same office holder was exacting fees from the regality of Arbroath, from which he had been excluded in 1359, suggesting the fee may have been partial compensation. Duncan, 'Laws of Malcolm MacKenneth', 257n. The coroner of Fife was entitled to a comparable fee in the 1650s. NRAS3215/Largo Papers/Bundle18/2A/18.

34 B. Webster (ed.), *The acts of David II, king of Scots, 1329–1371* (Edinburgh: Edinburgh UP, 1982), 312. For a later example (1565) see RMS III, no. 1609.

35 J. M. Brown, 'The exercise of power', in J. M. Brown (ed.), *Scottish society in the fifteenth century* (London: Edward Arnold, 1977), 56. Brown, *Bloodfeud*, 110–12. Forbes loyalty was also occasionally enforced by bonds of maintenance and manrent.

36 M. Brown, 'Scotland tamed? Kings and magnates in late medieval Scotland: a review of recent work', *Innes Review* 45 (1994), 120–46. Cam, *Liberties & communities*, 220. Those subsequently allowed to appoint 'special' or franchisal coroners were supposed to have explicit authority by charter. Lord Cooper, *Select Scottish cases of the thirteenth century* (Edinburgh: William Hodge, 1944), xliv–xlv.

37 Walker, *Legal history of Scotland* vol. 1, 215, 217.

38 Brown, 'Exercise of power', 64. Brown, *Bloodfeud*, 73. In Douglas and others v. Advocate (12 January 1748) the Court of Session recognised that subjects had had the right to grant heritable offices. Morison, *Dictionary*, 7699.

39 M. Brown, 'Earldom and kindred: the Lennox and its earls, 1200–1458', in S. Boardman and A. Ross (eds), *The exercise of power in medieval Scotland, c.1200–1500* (Dublin: Four Courts, 2003), 217.

40 NAS GD20/1/138. GD20/1/139-142, 147, 149. For other examples of 'listings' see ERS XIV, 517–18 (1513).

41 NAS GD160/128/4. The coronership of Strathearn had been in dispute since at least 1491: the issue was decided in favour of Drummond in the following year. GD160/4/14.

42 RPCS III, 491. NAS GD160/5/9.

43 RPCS V, 378–9. Rae, *Scottish frontier*, 145.

44 ATS I, 6; V, 82-3. NAS GD32/11/15. GD32/3/4. SC42/9/1 (27 October 1685). Sir John Murray eventually bought the coronership of Tweeddale from the heirs for 600 merks in 1595. J. W. Buchan and H. Paton (eds), *A history of Peeblesshire* 3 vols (Glasgow: Jackson, Wylie, 1925-7), vol. 2, 474, 510.

45 NAS RD1/20, part 2, f. 140b.

46 Mackenzie, *Observations*, 126. APS II, 330–1.

47 NAS JC1/2, f. 230v. Rae, *Scottish frontier*, 137. Murray, 'Administrators of church lands', 31.

DOI: 10.1057/9781137381071.0009

48   Pitcairn, *Criminal trials*, vol. 1, 106. APS II, 258–61. William Cunningham
      had sasine of '*officii* coronatoris mari feodi warde occidentalis de Stragriff et
      superioris warde barone de Renfrew' from his father in 1520. ERS XIV, 633.
      NAS GD148/31–9 and 180. Citations for treason were supposed to come not
      from a sheriff's officer like a coroner, but from a herald or pursevant. APS
      VIII, 480.

49   APS II, 258–61. NAS GD3/1/1/16/1. GD148/31–39. Eglinton's star was in the
      ascendant. Murray, 'Administrators of church lands', 35, 42.

50   I. H. Shearer (ed.), *Selected cases from Acta Dominorum Concilii et Sessionis,
      1532–1533* (Edinburgh: Stair Society, 1951), no. 69. Even in more peaceful times
      and places, coroners could be killed. NAS GD112/39/60/11.

51   A. Morison, *The Blackhalls of that ilk and Barra* (Aberdeen: New Spalding
      Club, 1905), 22. NAS GD148/31 (1484).

52   RPS I, 443. For other examples of grants to a husband and wife see R.
      Renwick (ed.), *Extracts from the records of the royal burgh of Lanark ... A.D.
      1150–1722* (Glasgow: Scottish Burgh Records Society, 1893), 361 and NAS
      RD1/20 part 2, f. 200. For the problems of lordships partitioned among
      females see Stringer, 'States, liberties and communities', 23.

53   For example W. M. Metcalfe (ed.), *Charters and documents relating to the
      burgh records of Paisley (1163-1665)...* (Paisley: Alexander Gardner, 1902), 24–5
      (April 1487). NAS CS7/23, ff. 139–139v (1553). APS V, 63, where the Earl of
      Sutherland was made sheriff and coroner there in 1633.

54   APS I, 492. MacQueen, *Common law*, 54, notes that such orders came at
      times of political instability and cannot be seen as a sign that justice was not
      administered effectively.

55   Dickinson (ed.), *Sheriff court book of Fife*, lxiii–lxiv. APS II, 17. Calderwood,
      Lords of council, 235. J. Stuart (ed.), *Extracts from the council register of the burgh
      of Aberdeen, 1398–1570* (Aberdeen: Scottish Burgh Records Society, 1844), 42.

56   Morison, *Blackhalls*, 17–18.

57   RPS II, 2081. GD32/20/8. Non-entry was failure of an heir to have himself
      served by proving title, allowing the superior (on a declarator) to enjoy the
      profits of the estate during the vacancy. Sinclair was a royal favourite and
      military counsellor who led the Scots to defeat at Solway Moss in 1542.

58   This is called *tosheadorach* in another grant. Skene, *Celtic Scotland*, vol. 3, 301.

59   ATS XII, 14–15, 23, 25, 373. NLS Ch. 12596. NAS GD135/1028. Cranston was
      put in charge of cavalry to pacify the eastern Borders in 1603, became deputy
      lieutenant of the region in 1605 and was made Lord Cranston in 1609. He
      meted out summary justice as well as arresting and presenting prisoners.
      Brown, *Bloodfeud*, 229. J. L. Hilson, 'The justice ayre', in *Yesterdays in a border
      parish* (Collected stories from the *Kelso Mail* c.1930), 9 [Edinburgh University
      Library Special Collections, Corson CC.Hilson 16].

DOI: 10.1057/9781137381071.0009

60 Perth & Kinross Archives B59/26/1/10. Hay described himself as one of the ordinary officers and crowners depute of the burgh. This document was witnessed by three named serjeants, showing that they were different from the coroner, at least in this time and place.

61 NAS CS233/S/1/30, 'Information for the sheriff of Bute and others heretors of the fewlands of Bute and their tennents, 1687'. Robert Jameson had transferred the office to Robert Ker of Baileachan.

62 NAS E21/73, f. 142r.

63 Calderwood, *Lords of council*, 202. For other examples of liability see pp. 77, 210, 323, 340. ERS XI, xlvii, 321. For another letter of remission addressed to multiple officers including coroners see NLS Adv. Ch. B.1320.

64 Hannay (ed.), *Lords of council, 1501–54*, 580, 617, 622, 623.

65 Webster (ed.), *The acts of David II*, 251, 379. APS I, 525. See also G. W. S. Barrow (ed.), *The charters of King David I: the written acts of David I King of Scots, 1124–53 and of his son Henry Earl of Northumberland, 1139–52* (Woodbridge: Boydell, 1999), 10–11.

66 Hannay (ed.), *Lords of council, 1501–54*, 620. RMS V, no. 389.

67 *Habakkuk Bisset*, vol. 3, 202–3. APS II, 36. APS I, 737. APS II, 177. Duncan, 'Laws of Malcolm MacKenneth', 256–7. Again, these acts coincide with periods of political crisis or instability and may not represent structural problems in the administration of justice. MacQueen, *Common law*, 54–5. For an example of a prosecution in 1503 for wrongful presentation of a man who had entered surety see Calderwood, *Lords of council*, 205–6. The deputy coroner was fined. On messengers see Sanderson, *Mary Stewart's people*, 135–48. I. D. Willock, *The origins and development of the jury in Scotland* (Edinburgh: Stair Society, 1966), 150–2.

68 APS II, 176.

69 Pinkerton (ed.), *Scottish poems*, 24. *Ancient Scottish poems*, 113.

70 L. J. Macfarlane, *William Elphinstone and the kingdom of Scotland, 1431–1514* (Aberdeen: Aberdeen UP, 1985), 113.

71 I owe this information to John Harrison. However, a decline in registration may simply mean that holders felt secure.

72 J. A. Clyde (ed.), *Hope's Major Practicks, 1608–1633* 2 vols (Edinburgh: Stair Society), 1937–8), V.3.15 and 18. ERS XI, xliii–xlv, xlvii. APS I, 571a; II, 23, 297, 331–2. *Atlas of Scottish history to 1707*, 211.

73 Wasser, 'Central criminal courts', 97, 173–5, 191. Wasser emphasises that, although the complaint may have arisen from a private individual, the prosecution was handled by royal officials. APS III, 458–9. J. Louthian, *The form of process before the court of justiciary in Scotland* (Edinburgh: R. Fleming, 1732), esp. 15–18, 210–11. *A treatise on the duties and office of a messenger at arms* [no title page: manuscript description, 1801], 2–3 [NLS ABA.1.78.405].

DOI: 10.1057/9781137381071.0009

This would seem to be Robert Thomson, *A treatise of the office of a messenger* (Edinburgh: T. & W. Ruddiman, 1753).

74   H. M. Dingwall, *Late seventeenth-century Edinburgh: a demographic study* (Aldershot: Scolar, 1994), 216.

75   Stevenson, 'Officers of arms', 41–66. C. Burnett, 'Early officers of arms in Scotland', *Review of Scottish Culture* 9 (1995–6), 3–13. Godfrey, *Civil justice*, 228.

76   A. Frazer, *A treatise on the office of messenger, as now practised in Scotland* (Edinburgh: J. Ruthven, 1815), 1. APS III, 449, 457, 460. *A treatise on the duties and office of a messenger at arms*, 7–8.

77   Frazer, *A treatise on the office of messenger*, 11. Burnett, 'Early officers of arms', 8–9.

78   Stevenson, 'Officers of arms', 57, 61–4.

79   NAS GD52/69 (1556/7).

80   Burnett, 'Early officers of arms', 6. One example among many is 'A proclamation anent the Sumptuary Act, 1684'.

81   NAS E21/73, ff. 109r, 142r.

82   Wasser, 'Central criminal courts', 173.

83   *A treatise on the duties and office of a messenger at arms*, 3–7. Darling, *The powers and duties of messengers-at-arms*, 12–15. Burnett, 'Early officers of arms', 8–9. M. H. B. Sanderson, *A kindly place: living in sixteenth-century Scotland* (East Linton: Tuckwell, 2002), 64–5. Stevenson, 'Officers of arms', 63–4. R. Robertson, *The history and present constitution of the Sheriff Courts of Scotland* (Glasgow: J. Maclehose, 1863), 5. Rebellion against the crown was the foundation for execution of caption in cases of civil debt. A debtor who failed to pay when duly charged with letters of horning became a rebel when publicly denounced in the king's name. *Balfour's Practicks*, 566. J. Dalrymple, *An essay towards a general history of feudal property in Great Britain* (1757. 4th edition. London: A. Millar, 1759), 59–60. NAS E21/72, f. 48v. Ross, *Lectures on diligence*, vol. 1, 273–9. D. Murray, *Early burgh organization in Scotland* 2 vols (Glasgow, 1924, 1932), vol. 2, 515–17. I. Treiman, 'Escaping the creditor in the Middle Ages', *Law Quarterly Review* 43 (1927), 230–7. Goodare, *Government of Scotland*, 178. Ford, *Law and opinion*, 161. Messengers have since 1510 been subject to the control of the Lord Lyon King of Arms.

84   Mackenzie, *Observations*, 91, 92, 126. A generation later still, Louthian, *Court of justiciary*, 287–97, wrote that porteous rolls were still in use on circuit courts, but that the official charged with attaching and other duties was a 'macer' or usher. 'The Justice Court has three Macers, tho' they are not restricted to that Number. They have their Commissions from the Justice General, ... and each of them has a Mace and black gown. The Macer's chief Business is, to execute all Indictments, Criminal Letters, &c, citing of Assysers and Witnesses against and for Pannels [the accused], executing Warrants for Imprisonment, waiting on the Judges all the Dyets of Court,

DOI: 10.1057/9781137381071.0009

bringing out and returning Pannels from and to Prison, to inclose Juries, and attend upon them while they are drawing up their Verdict, and repeats after the Clerk all Sentences that are not Capital.' *Ibid.*, 7.

85 Pinkerton, *History*, vol. 2, 404. Glassford, *Scottish courts of law*, 47, says that 'the office of the justiciar was cut down by James V'.

86 Goodare, *State and society*, 148. Wasser, 'Central criminal courts', 96–7. A. Blakeway, 'Regency in sixteenth-century Scotland' (Cambridge University Ph.D., 2009), 174, 176, 184, 187–8.

87 APS III, 458–9. R. K. Hannay, 'The office of the justice clerk', in H. L. MacQueen (ed.), *The College of Justice* (Edinburgh: Scottish Academic Press, 1990), 338–9. G. Donaldson, *Scotland: James V to James VII* (Edinburgh: Oliver & Boyd, 1971), 5, 56. Walker, *Legal history of Scotland* vol. 3, 132–3, 426. J. Findlay, *All manner of people: the history of the justices of the peace in Scotland* (Edinburgh: Saltire Society, 2000), 21. J. Glassford, *Remarks on the constitution and procedure of the Scottish courts of law* (Edinburgh: Archibald Constable, 1812), 116. Commissions of justiciary were suspended in the Borders after 1590 in favour of hearings in Edinburgh. Armstrong, *History of Liddesdale*, 10. D. L. W. Tough, *The last years of a frontier: a history of the Borders during the reign of Elizabeth 1* (Oxford: Clarendon Press, 1928), 171.

88 NAS E21/70, ff. 160r–160v, 191v. E21/71 ff. 76r, 108r–109r, 110v, 115v. E21/72, ff. 45r–45v, 48v, 52v, 54r, 55r, 61r, 62v. E21/73, ff. 109r, 112v, 114v, 130r. E21/74, 106v. E21/76, f. 118v, 191r–195r, 201v, 232r. I owe these references to Michael Wasser.

89 RPCS 2nd series II, 436–7. He retained the office of justiciar of Argyll and the Isles. The Earls of Argyll had held the position since 1514. Davies, 'Scottish legal system', 147. Walker, *Scottish legal system*, 82, states that the office was not united with that of Lord President of the Court of Session until 1836.

90 'Account of a journey into Scotland, 1629', *The Scottish Antiquary* 10 (1895), 31.

91 C. S. Terry (ed.), *The Cromwellian union … 1651–1652* (Edinburgh: Scottish History Society, 1902), 181n.

92 *The laws and customs of Scotland in matters criminal*, in *The works of that eminent and learned lawyer, Sir George Mackenzie of Rosehaugh, advocate …* (Edinburgh: J. Glen, 1678), 428.

93 Cameron, *Justiciary records*, 7, 26, 48, 71, 90.

94 *Ibid.*, 7, 36, 48, 63–4, 71, 90. For example, in 1674–5 the procurator was Duncan Fisher, the crowner (of Knapdale) Donald M'Neill. The M'Neills had been *toiseachdors* or coroners there since at least the fifteenth century. Munro, *Lords of the Isles*, lii and no. 33.

95 NAS CC12/1/1, f. 82 (12 March 1667). Robert Jameson had been the crowner in the 1650s and had owed Robert Ker 3,240 merks. Ker was nominated executor because he was Jameson's creditor. The level of exaction on Bute was more onerous than on the Isle of Arran, where the same payment was due only from each farm toun or settlement. Martin, *Description*, 224.

DOI: 10.1057/9781137381071.0009

Coroners elsewhere got payments from feuars. NRAS3215/Largo Papers/ Bundle18/2A/18 (Dairsie, 1652). This document persistently renders coronership as 'crimmary' of justice-ayre courts. A feu or fee farm was a perpetual lease, usually granted by a superior in exchange for a single large capital sum or *grassum* and small subsequent fixed periodical payments (*reddendo*); it was *dominium utile*, the estate held by the lowest vassal.

96   NAS CS233/S/1/30, 'Information for Mr John Steuart of Ascog advocate against the Shirref of Bute and others, 1687'.

97   F. J. Grant (ed.), *The faculty of advocates in Scotland, 1532–1943* (Edinburgh: Scottish Record Society, 1944), 200–1.

98   NAS GD112/56/27. CS233/S/1/30 (1687), 'Information for the sheriff of Bute and others heretors of the fewlands of Bute and their tennents, 1687'. CS233/S/1/34. GD124/15/211, 'Letter to John Steuart of Ascog, advocate from J. Steuart about uplifting of crownry'. APS VIII, 379. *Historical notices*, 624, 847. Mackenzie, *Observations*, 126, presumably had the Bute spat in mind. Ascog subsequently transferred the office to the sheriff, possibly persuaded that if he wanted the perquisites he had to keep men to arrest criminals. Forbes, *Justices of peace*, pt. 2, 11.

99   NAS CS233/S/1/30, 'Information for the sheriff of Bute and others heretors of the fewlands of Bute and their tennents, 1687'. I have compared named coroners holding after 1532 with the printed lists of advocates and writers to the signet, and with Finlay's *Men of law*. Nor did Scottish Justices of Peace need formal legal training. Tait, *Justices of the peace*, 172. J. Chynoweth, *Tudor Cornwall* (Stroud: Tempus Publishing, 2002), 201.

100   Frazer, *A treatise on the office of messenger*, 1. Candidates were examined before the Lord Lyon or his depute in Edinburgh from 1587. *Ibid.*, 2. *A treatise on the duties and office of a messenger at arms*, 9–10. Stevenson, 'Officers of arms', 61–2.

101   Walker, *Legal history of Scotland*, vol. 1, 216, 221. The justice clerk may have had legal training from an early date and certainly did from the sixteenth century. *Ibid.*, 222–3. Walker, *Legal history of Scotland*, vol. 3, 133.

102   Bellamy, *Crime and public order*, 95. Smyth, *De republica Anglorum*, 2:21 (p. 72).

103   Stevenson, 'Officers of arms', 50–1.

104   Thomson, *A treatise of the office of a messenger*, 2.

105   Forbes, *Justices of peace*, pt. 2, 11.

106   NAS GD1/19/18. GD1/19/20. Martin, *Western Islands*, 224. In a case of 1742, part of the rents on the island were described as the 'crownary'. NRAS332/C3/441.

107   J. Imrie (ed.), *The justiciary records of Argyll and the Isles, 1664–1742, vol.* 2 Stair Society 25 (Edinburgh, 1969), 226.

108   NAS RH15/19/61.

DOI: 10.1057/9781137381071.0009

109 APS VIII, 390a. This may have been a political move since Dunbar was later elected as a representative to the Convention of Estates called by William III in 1689. He appears in parliament again in 1700. A. Agnew, *The hereditary sheriffs of Galloway* 2 vols (Edinburgh: D. Douglas, 1893), vol. 2, 153, 177.

110 8 & 9 Anne c. 16. Hutcheson, *Justice of peace*, vol. 1, 83n. Louthian, *Court of justiciary*, 280–7. Hume, *Commentaries*, vol. 2, 24–6. Davies, 'Scottish legal system', 146–51.

111 Davies, 'Scottish legal system', 147–8, 150. There was no grand jury system for indictments of the kind found in England. The advocate-depute (who originated in the late sixteenth century) fulfilled some of its functions. R. S. Shiels, 'Crown Counsel: from Sir Archibald Alison to Lord Brand', in J. Chalmers, F. Leverick and L. Farmer (eds), *Essays in criminal law in honour of Sir Gerald Gordon QC* (Edinburgh: Edinburgh UP, 2010), 287–8.

112 20 Geo. II, c.50. This may be why Smith, 'Criminal procedure', 427, says the office only became obsolete in the eighteenth century.

113 Hutcheson, *Justice of peace*, vol. 1, 4n. NLS Ry.III.a.16(315), 'The petition and claim of Thomas [Lyon], Earl of Strathmore, heritable Constable of the Burghs of Forfar and Kinghorn, and heritable Coroner within the Bounds of the Shires of Forfar and Kincardine ... Nov. 10 1747'. The printed submission is annotated with an abstract of judgement in respect of the coronership given on 4 February 1748, stating that the office did not fall under the Heritable Jurisdictions Act. A later judgement (25 February 1748) found in favour of Strathmore in respect of the constableship of Forfar, but judged his title to that of Kinghorn defective. See also NLS Ry.III.a.16(316). The only other hereditary coroners to claim were John, Earl of Hopetoun for the regality of St Andrews south of the Forth, William, Earl of Sutherland for the shire of Sutherland and Roger MacNeill of Taynish for the lands of Gigha, his offices held from the Duke of Argyll. *Roll or list of the claims entered in the Court of Session in Scotland in pursuance of an ... Act for abolishing heretable jurisdictions* (London, 1748).

114 Hume, *Commentaries*, vol. 2, 24n.

115 NAS GD64/1/4/16. For an earlier renunciation of the same office, see GD64/1/3/1 (1696).

116 NAS GD1/221/96. Securing payment had been a problem for holders since at least the 1620s. NLS Ch. 7972–81.

117 There is no mention of coroners in the section on arresting delinquents in Clark, *Office of sheriff*, 12–14. For a comparable shift from traditional, community-related officers to newer and perhaps more centrally answerable ones, see M. Goldie, 'The unacknowledged republic: officeholding in early

DOI: 10.1057/9781137381071.0009

modern England', in T. Harris (ed.), *The politics of the excluded, c.1500–1850* (Basingstoke: Palgrave Macmillan, 2001), 153–94.

118 D. M. Walker (ed.), *The institutions of the law of Scotland ... by James, Viscount of Stair ... 1693* (Edinburgh: Stair Society, 1981), 60 ('Dedication to the king').

119 NLS Ry.III.a.16(316).

120 Hutcheson, *Justice of peace*, vol. 1, 4.

DOI: 10.1057/9781137381071.0009

5

Regional and National Histories: Similarities and Differences between the Coroners of Northern Britain

Abstract: *This chapter teases out some intriguing similarities in the functions of coroners in northern England, Wales and Scotland up to the sixteenth century, arguing that certain regions of Britain shared important characteristics, which made them different from southern and Midland England. In particular, peace-keeping in the north and west of Britain involved kin and lords, whereas in the south and east responsibility for local enforcement lay with communities. The coroners of the north of England and Scotland shared broadly similar administrative, judicial and military (including law and order) duties until the sixteenth century. Only then did all English coroners come to deal primarily with sudden, suspicious or unexplained deaths.*

Houston, R. A. *The Coroners of Northern Britain c. 1300–1700*. Basingstoke: Palgrave Macmillan, 2014.
DOI: 10.1057/9781137381071.0010.

For all the differences at a national level, there are also intriguing similarities between the status and functions of coroners in Scotland and the north of England, especially in liberties and franchises, which speak to the continued importance of regional as much as national political, social and legal structures in Britain. Blackstone regarded the offices of coroner and sheriff as incompatible in England because the former was a crown officer, one of whose tasks was to investigate accusations of partiality against the latter.[1] Coroners kept the pleas of the crown rather than holding them and they kept and transmitted records of the administration of justice rather than hearing and determining causes.[2] This was probably true of much of England by the end of the Middle Ages, but in late-medieval Northumberland coroners indicted offenders and in Co. Durham and Cheshire they 'were very much the partners of the sheriff – and his equals – in the administration of the criminal law', including before sessions of gaol delivery.[3] The two were meant to go hand-in-hand on behalf of the king and in fourteenth-century Northumberland coroners were active beyond holding inquests. There and possibly in Newcastle-upon-Tyne too they took the place of the sheriff's tourn (twice-yearly circuit court) for felonies, the local 'peace' system relying on presentments made to coroners; they rather than the sheriff prepared the calendar of offences for trial at the gaol delivery sessions of the general eyre. In the north-east, coroners may have carried on judicial functions that had been curbed elsewhere by Chapter 24 of Magna Carta.[4]

Sixteenth-century sheriffs of Northumberland were seldom called to account for their income and expenditure, and were generally allowed to pay a flat fee to the Exchequer.[5] However, relations between sheriffs and local magnates were as important as those with the crown in the north of England and those for counties Palatine (Chester, Durham and Lancaster) had separate accounting procedures.[6] For example, sheriffs *might* pay in to the bishop of Durham the profits from tourns and other courts, the value of goods forfeited or distrained to pay fines levied or for the goods of a hanged person, a suicide or a runaway. However, the section for sheriffs is usually blank in the sixteenth-century Bishopric's 'Books of Great Receipt', the sheriff being a relatively free agent.[7] The reason was less their power and independence than the opposite. The standard work on the King's Council in the North states that 'sheriffs beyond the Trent had but small force' as jurisdiction was often parcelled out in manors, baronies and honours.[8] Sheriffs nationally became controversial figures because of their involvement in the collection of Ship Money in the 1630s. After the Restoration, the growing importance of JPs as county authorities marginalised sheriffs to the status of semi-detached agents of county government.[9] The job became unpopular and could be used to punish insubordinate gentry rather than recognise and reward the well-affected, meaning that the time and expense involved made holders and their allies little inclined to act expeditiously on behalf of the crown.[10]

Both in the Middle Ages and after, the franchisal coroners of Northumberland and Durham had more extensive powers than their counterparts elsewhere in England, making them closer to their Scottish namesakes. The bishop of Durham appointed coroners from the thirteenth century until 1836, along with other officers who were part of his financial administration. Medieval Episcopal commissions of array included coroners and they participated actively in the prosecution of cross-Border crime.[11] They took custody of vacant estates and victualled for bishops, acting as bailiffs in courts, outlawing and capturing fugitives, aiding in raising fencible men as well as investigating treasure-trove and suspicious deaths.[12] The four for the wards of the bishopric of Durham (Chester, Darlington, Easington and Stockton plus ones at Norham, Bedlington and the wapentake of Sadberge) also kept rentals of freehold dues for the bishop's estates.[13] The coroners of the medieval palatinate derived their fees either from levies on the vills, as did some Scottish coroners and earlier Welsh serjeants of the peace, or sometimes directly from the bishop.[14] They were every inch the bishop's men and, both before and after the Black Death, he used the officers of local government to support his estate policy, his coroners arresting or distraining defaulting tenants.[15] Some Welsh coroners did the same jobs, notably in the early-Tudor Englishry of Gower. The coroner there collected the rents of the Lord of Gower Anglicana (including handling guardianships) and also bringing in the perquisites of the shire court of the lordship; on occasion he performed the tasks that might elsewhere be done by an escheator. Chosen by the lord from three nominees of the shire court, he held office by the year, assisted by two bailiffs.[16]

It was rare for English coroners to have attached tenures though some thirteenth-century serjeanty tenants in Northumberland resembled certain Scottish coroners in holding land in return for making arrests and seizures as (named) coroners.[17] Scotland had, however, no serjeanty tenures in the English sense of officers holding their positions by that tenure.

DOI: 10.1057/9781137381071.0010

Men sometimes held land in exchange for acting as coroner and some may have persisted long enough for place names such as 'Crowdarland' ('Crownerland') in West Lothian to stick. For Scotland's hereditary holders, however, land and title could be disponed separately, suggesting that the arrangement for their subordinates was a convenience rather than a tenure in its own right.[18] A heritable coronership, that could be given and disponed by charter, did not require sasine in the same way as land because it was *merum jus incorporeum* (merely an incorporeal right).[19]

These distinctive roles and conditions of work existed elsewhere in the north of England. Coroners on the isle of Man, appointed by the governor and with duties comparable with a mainland sheriff, received a share of the forfeited goods of felons and suicides, as well as an annual contribution from the lord's tenants.[20] This made them different from most mainland coroners and closer to the Durham coroners or to the coroner and governor of the regality of Kilbride or Arran noted earlier; they may have been the successors of the *toíseachdeors* of the island's six medieval 'sheadings' or administrative units.[21] As the chief executive legal officer, the early modern Manx coroner empanelled all sorts of juries, arrested and presented offenders and made a quarterly 'general search' for stolen goods; like Scottish coroners he bore arms. Early sixteenth-century manorial records from the island show fines for disobedience to the lord's coroner.[22] As in the palatinate of Durham, coroners could appoint deputies and the Manx coroner's was the parish lockman or constable. The *moar, maor* or mair on Man was a lowlier officer than his Scottish equivalent: an elected bailiff of the lord, charged with collecting forfeitures and other feudal accidents, though the coroner too performed tasks like presenting sexual miscreants for whipping.[23] Annual accounts from the years 1511–15 show that there were separate *moars* and coroners for each parish or group of parishes and that each accounted separately for income.[24] Again, in these franchisal appointments we find greater similarities with Scotland prior to the mid-eighteenth century, possibly thanks to enduring Norse influence on Man even after its takeover by English lords in 1406. An important contrast is that the coroners, lockmen and moars held office for one year only, possibly in rotation – as with other local functionaries.[25]

The north of England was rich in franchisal coroners, appointed by a lord whose charter allowed it. These semi-private coroners had to act both *ad facienda negotia Regis* (to do the king's business), but also as a sort of lord's steward *pro voluntate sua facere* (doing his will).[26] In jurisdictions

DOI: 10.1057/9781137381071.0010

like Holderness in Yorkshire, the coroner was an appointee of the lord of the manor and held office at his pleasure.[27] In the case of the lordship of Egremont and honour of Cockermouth in Cumberland, the coroner was also an appointee of the lord.[28] The coroner of the wapentake of Sadberge in Co. Durham was a hereditary office tied to a holding of land by serjeanty; as in the four wards of Durham itself the coroner was an appointee and employee of the bishop, who accounted for forfeitures due at the palatinate Exchequer.[29] In 1516, a Darlington coroner found himself imprisoned for arrears due to his bishop and Durham coroners as a whole were very much bishop's men.[30] In passing one might note that the fourteenth-century prior's court had its own coroner, who may have sat with the court's president, though his main function seems to have been selecting and swearing in juries. He was important either in ensuring fair play or in weighting justice towards the lord.[31] Elsewhere too there were lesser coroners who had duties comparable with stewards or overseers. The manor of Ashby de la Zouche in Leicestershire had 'Coroners of the Markett' in the seventeenth century.[32] In 1636, Queen Henrietta Maria conferred on Thomas Cholmeley of Carlisle, gentleman, the office, during pleasure, of guardian and bailiff, coroner and clerk of the market within her possessions in Cumberland with a fee of 10s. a year.[33]

The hands of lords temporal as well as spiritual fell on the shoulders of the king's men in the north, for even private coroners swore an oath to the crown. In 1523, the fifth Earl of Northumberland wrote to the eleventh Lord Clifford about matters in Cumberland. 'I am informed that all the crowners in this country is bound to certifie to the kinges counsaile of all the owtlawries in this country; wherefore it is good your lordshipp speake with the crowners in your partes for your frends and servantes for suites, for if the outlawries be delivered it will sure come to a money matter: if they be staid and not delivered and the parties agreed it shall doe no hurt.'[34] With echoes of the Scottish preference for assythement (and of the Welsh for *galanas* or the Irish for *eric*), the Earl advocated informal accommodations that kept the hand of outsiders from local and regional disputes.[35] Northumberland hoped that pressure from superiors could be brought to bear. Yet at the same time, his statement acknowledged that the coroner was the king's man and it offered a stark recognition of his mission in a part of the realm often seen as a prominent sign that early Tudor England was 'a federation of noble fiefdoms'.[36]

DOI: 10.1057/9781137381071.0010

A difficult case from the Civil War period shows the dynamics of rivalries. John Boulton of Highside in Embleton (Cumberland) hanged himself on Sunday 23 September 1648. A jury under the direction of Mr John Lamplugh, coroner to the Earl of Northumberland, entered a verdict of *felo de se*. A dispute then arose over who should receive the goods and chattels of the deceased. The lord of the manor of Cockermouth (the Earl) and the high sheriff of Cumberland contested the jurisdiction over such matters within the liberty of Cockermouth. A detailed inventory revealed that Boulton's personal estate was worth £58-6-8, including rentals from the Earl and he also rented 'ane tenement of Mr Braythwaite'; he owed debts of £7-15-6. Amongst the depositions relating to the case, the top sheet is a letter from the Earl's bailiff, Edmond Grainger, explaining to his master why the high sheriff (John Barwis, Esq.) had taken custody of livestock belonging to Boulton and how the sheriff laid claim to the forfeiture, 'locked upp the barne doores' and 'carryed away' certain oxen. Shortly afterwards, Grainger 'alsow locked upp the barne doores' and seized the remaining livestock 'for the Lord's use'. The sheriff then sent his 'baylifes with his warrant' to take the remaining livestock from Grainger, threatening to 'send Troups' if Grainger refused (as he did).

Well out of his political depth, Grainger sent a deferential letter to the sheriff, who was, nevertheless, true to his word and dispatched his own bailiff, William Young, with 'fower troups armed' to repossess the goods on 8 December. Grainger went on to complain that the sheriff, 'broke off the locks from the barne doores which I had sett on, myselfe being present, and after the locks were of[f] I did resist the Sheriff and would not suffer him to enter the barnes till by force hee pulled me away, soe his servants threshed out all the corne, solde the hay and disposed of other goods which would amont to a great sume'. The notes from the coroner's inquest show that the sheriff ultimately accepted the Earl's 'Crowne charter' to collect the forfeited goods and repaid £40-16-8 (the sheriff deducted £12-3-6 for himself and Grainger received £5-6-6).[37]

The area generated further contests over rights. In the autumn of 1673, Sir John Lowther requested that William Smith report upon an inquest undertaken near Cockermouth by John Lamplugh, which had caused controversy. The case concerned a man surnamed Todhunter, found drowned at Patterdale. On the surface, there was a disagreement within the coroner's jury as to whether Todhunter had committed suicide. On closer scrutiny, however, it again turns out to be equally a jurisdictional dispute. Lamplugh was anxious that Todhunter should be a *felo de se*

DOI: 10.1057/9781137381071.0010

and so seized his assets, claiming that he had a deputation to take all the goods from those found *felo de se* in the manor where the body had turned up. Patterdale was on the fringes of the Earl of Northumberland's lands and Lamplugh's right to hold the inquest was in question, as much as Todhunter's mode of death.[38]

Lamplugh was a franchisal coroner appointed by his lord. The power of the great nobility endured in the north and, because county coroners were elected, they remained political pawns into the eighteenth century. Coroners could be malleable, political creatures whose elections came to have a political edge, with local magnates keen to have their man chosen as a way of announcing their political style and the extent of their influence. Elections were a bell-wether of county politics. In 1763, Henry Curwen wrote to the Duke of Portland to ask if he wanted to run an opponent for county coroner of Cumberland against the candidate supported by Sir James Lowther. Curwen thought that Lowther was 'grasping at the minutest things that may extend his rule over the county' and suggested that the election would allow the Duke to try the broader political spirit of the county for parliamentary contests.[39]

Delegation of function, common in Scotland, was also a feature of some northern English coronerships of the thirteenth century, notably when the holder in fee was a woman.[40] Some later northern boroughs also tolerated delegation. At Liverpool, payments went to the mayor as official coroner, leaving the outgoing mayor and later the bailiffs to do an arduous job unpaid.[41] In other boroughs too coroners were delegated or elected members of the common council for a year at a time.[42] This went against the usual English, Welsh and Irish rule that 'the coroner go in proper person to do his office ... he can not make a deputy', a provision not changed until 1843.[43] Perhaps because of the common Anglo-Norman origins of burgh/borough law, this arrangement (like that in London prior to 1478) more closely resembles certain Scottish royal burghs, like Aberdeen, Edinburgh, Inverness and Perth, where the crown granted the right to the council to elect a provost, who became *ex officio* (by virtue of his office) coroner (or sheriff and coroner) with the bailies as deputies.[44] At Perth, for example, the provost and two bailies took out notarial instruments, before the Justiciar's Court sitting there on 4 February 1510, confirming their right '*officii coronatour vulgariter crownarschip*' (to the office of coroner, commonly known as the crownership).[45]

The final similarity lies in the way the functions of coroners in both Scotland and England gradually changed by erosion or substitution,

DOI: 10.1057/9781137381071.0010

rather than outright reform. The change happened rather later in Scotland and the mechanism was different. English coroners' roles mutated thanks primarily to the strengthening of other officers of the crown, the JPs acting under commission, and the decline of eyres and sheriffs. Coroners nevertheless retained important roles in a system of government that was more participative than in Scotland. The Scottish coroner became marginalised by developments in justice, brought about by the crown and lawyers in the central courts, which gradually rendered his job superfluous.

This does not mean that the office and its context were identical across northern and western Britain. Away from the palatinate of Durham, parts of Northumberland, the isle of Man and the Welsh Marches, English franchisal coroners more closely resemble stewards than do Scottish coroners (or *toíseachdeors*) – even those appointed by the nobility, whose obligation to serve royal justice-eyres remained paramount.[46] Over a longer time-span the contrasts became stronger than the similarities, even in the north of England. In Cheshire, for example, coroners had been charged with apprehending accused felons in the thirteenth century, but by the fourteenth they had to be given special powers as *custodes pacis*.[47] Serjeants in the north of England increasingly dealt with the pleas of the crown, becoming separated from the sheriff and in time (like coroners) under direct crown control.[48] A Star Chamber bill of complaint of Henry VIII's reign accused Henry Hookenhall, coroner of the hundred or wapentake of the Wirral, of misdemeanours in office over an alleged murder. One point of the bill claimed he had refused to allow 'Wylliam Clayton, one of the kynges serjaunts in those quarters, to give evidence for the kyng nor to chalenge the enquest'. More detail followed. 'The seid serjaunt requyryd the seid coroner that he myght go into the house where the enquest was put after yt was chargyd, and he wolde not suffer hym to cum where they were to enforme them of no thyng uppon the kynges behalff; nor wolde he suffer hym to here and be at the takyng of the verdyt for the kyng, to thentent to make it in forme of the lawe, but did cause hym to avoyde the church where he satt, and locked the church dore to kepe hym oute, that he sholde not be prevy to the same'.[49] Things were changing in the north even as these events unfolded. What is sometimes termed 'Welsh legislation' of 1534–43 formalised coroners in Cheshire.[50] The contemporary statute of franchises (27 Hen. VIII, c. 24) insisting that justice be in the king's name alone, weakened the judicial scope of Durham

DOI: 10.1057/9781137381071.0010

coroners, who in time became more like their counterparts elsewhere.[51] Thus by Elizabethan times 'the great palatinate jurisdictions are best seen as local expressions of royal authority rather than delegations of it.'[52] Would further investigation show that franchisal coroners outside the north behaved in similar ways?

The other main difference between the countries lay in the social standing of coroners. There was no formal property qualification for Scottish coroners (or Justices of Peace for that matter), but they were closer in status to English escheators than coroners.[53] The enduringly high status of Scottish holders shows the strong connection between royal justice and the leadership of local communities. A Scottish coroner held an honourable office carrying far more prestige than in England, a contrast especially clear in the sixteenth and seventeenth centuries.

There was a low property qualification for early modern and later English coroners, though medieval holders may have been more socially elevated, as Blackstone suggested, and thus closer in status to their Scottish equivalents. Early modern English coroners were socially more like the earlier hundred serjeants they had replaced after 1194.[54] In late-fifteenth-century Cheshire most coroners were from the ranks of the middle gentry.[55] Even in the palatinate of Durham, where they were usually chosen from among tenants-in-chief, the social status of late-medieval coroners was lower than was usual among Scottish titular holders.[56] Some earlier Durham coroners were familiars or servants of the bishop, yet the overall sense is of *mediocres*.[57] Meanwhile in Westmorland during the 1440s the coroners were allegedly 'the meynyall men' of the maverick under-sheriff.[58] Only in Tynedale, where Alexander III had once appointed coroners, was the status higher; there were other contrasts in that men here were generally elected to the office for short periods during the thirteenth and fourteenth centuries.[59]

Notes

1   Blackstone, *Commentaries*, I.9.II [vol. 1, 337]. M. Dalton, *Officium Vicecomitum. The office and authority of sheriffs: gathered out of the statutes and books of the common laws of this kingdom* (1623. London, Richard Atkins, and Edward Atkins, 1682), 46–93. J. Impey, *The new instructor clericalis ... also, the office of sheriff* (London: W. Strahan and W. Woodfall, 1782), 97–9. Hunnisett, *Medieval coroner*, 2–3, 86–9.

DOI: 10.1057/9781137381071.0010

2   F. Pollock and F. W. Maitland, *The history of English law before the time of Edward I* 2 vols (1895. 2nd edition. Cambridge: Cambridge UP, 1898–1911), vol. 1, 534.

3   F. Pulton, *An abstract of all the penal statutes which be generall, in force and use* ... (London: C. Barker, 1592), 'Coroners', heading 20. The reference is to the two head coroners of Cheshire. T. Thornton, *Cheshire and the Tudor state, 1480–1560* (Woodbridge: Boydell, 2000), 132n. Jacob, *Law-dictionary*, 'coroner'. C. J. Neville, ' "The bishop's ministers": the office of coroner in late medieval Durham', *Florilegium* 18, 2 (2001), 52. Summerson, 'Peacekeepers', 58–9, 75. There is also some evidence of coroners acting as keepers of the peace and sitting as justices of gaol delivery in late-thirteenth- and early-fourteenth-century East Anglia. A. Musson, *Public order and law enforcement: the local administration of criminal justice, 1294–1350* (Woodbridge: Boydell, 1996), 153–4.

4   Summerson, 'Peacekeepers', 58–9, 67. Harding, 'Keepers of the peace', 95. Morris, *Frankpledge*, 48–55. M. F. Moore, *The lands of the Scottish king in England: the honour of Huntingdon, the liberty of Tyndale and the honour of Penrith* (London: George Allen & Unwin, 1915). Hunnisett, *Medieval coroner*, 5. At a Newcastle eyre in 1255/6 the sheriff and coroner were the same man. *Three early assize rolls for the county of Northumberland, sæc XIII* (Durham: Surtees Society, 1891), 68. NA JUST 3/54/3 mm 1–4. There is evidence of a tourn for the borough of Newcastle, which was administered as a shire, in 1489 – though legislation of 1461 had transferred many criminal functions to JPs. NA KB 9/382, mm 6–7. J. Armstrong, 'Local conflict in the Anglo-Scottish borderlands, c.1399–1488', (Cambridge University Ph.D., 2007). C. Fraser, 'Justice in Northumberland in the Middle Ages', *Tyne and Tweed* 59 (2005), 3–13. Dawson, *Lay judges*, 182–3, 189–91.

5   Dalton, *Officium Vicecomitum*, 474–81. M. C. Noonkester, 'Dissolution of the monasteries and the decline of the sheriff', *Sixteenth Century Journal* 23, 4 (1992), 688. C. H. Hunter Blair, 'The sheriffs of Northumberland', *Archaeologia Aeliana* 20, 21, 22 (1942, 1943, 1944), 11–90, 1–92, 22–82. For Exchequer revenues and accounting procedures see G. Gilbert, *A treatise on the Court of Exchequer* ... (London: H. Lintot, 1758), esp. Ch. 7–10.

6   Dalton, *Officium Vicecomitum*, 504. Impey, *Office of sheriff*, 104–8.

7   Durham University Archives CCB/B/28/23–25. CCB/B/29/26–33.

8   Moore, *Lands of the Scottish king in England*. Hunnisett, *Medieval coroner*, 5. R. R. Reid, *The King's council in the north* (London: Longmans, 1921), 8. S. G. Ellis, 'Tudor Northumberland: British history in an English county', in S. J. Connolly (ed.), *Kingdoms United? Great Britain and Ireland since 1500: integration and diversity* (Dublin: Four Courts, 1999), 35. Clayton, *Administration of Chester*, 172–81, offers a more nuanced picture of sheriffs' duties in an area with multiple franchises.

9   Dalton, *Officium Vicecomitum*, 81. J. Mather, 'The civil war sheriff: his person and office', *Albion* 13, 3 (1981), 242–61. J. Innes, 'Governing diverse societies',

DOI: 10.1057/9781137381071.0010

in P. Langford (ed.), *The eighteenth century* (Oxford: Oxford UP, 2002), 103–4, 107. Medieval English coroners had experienced the same sidelining, thanks to the growth of JPs. Hunnisett, *Medieval coroner*, 191–9.

10   E. Marcotte, 'Shrieval administration of ship money in Cheshire, 1637: limitations of early Stuart governance', *Bulletin of the John Rylands Library of Manchester* 58 (1975–6), 137–72. J. M. Rosenheim, 'Party organization at the local level: the Norfolk Sheriff's subscription of 1676', *Historical Journal* 29 (1986), 713–22.

11   Neville, 'Office of coroner', 52–3. The commission was Episcopal not royal, but the troops were for royal service. Redesdale was also a recruiting unit and its lords had the powers of sheriffs to try pleas including those of the crown. R. Robson, *The English highland clans: Tudor responses to a medieval problem* (Edinburgh: John Donald, 1989), 7. M. Prestwich, ' "*Tam infra libertates quam extra*": liberties and military recruitment', in M. Prestwich (ed.), *Liberties and identities in the medieval British Isles* (Woodbridge: Boydell, 2008), 111–19. M. James, *Family, lineage, and civil society: a study of society, politics, and mentality in the Durham region, 1500–1640* (Oxford: Oxford UP, 1974), 42–4. Meikle, *British frontier*, 95. Durham was not in the marches, but the bishop had extensive lands in Northumberland.

12   K. Emsley and C. M. Fraser, *The courts of the palatinate of Durham from earliest times to 1971* (Newcastle: Pattinson and sons, 1984), 15. C. M. Fraser and K. Emsley, 'Justice in north east England, 1256–1356', *The American Journal of Legal History* 15 (1971), 163–85. P. L. Larson, 'Local law courts in late medieval Durham', in C. D. Liddy and R. H. Britnell (eds), *North-east England in the later Middle Ages* (Woodbridge: Boydell, 2005), 101. C. M. Fraser (ed.), *Durham quarter sessions rolls, 1471–1625* Surtees Society 199 (Newcastle, 1991), 99, 105–6, 150. Stewart-Brown, *Serjeants of the peace*, 69–72, believes military serjeants of the peace were quite different from 'the more normal type', but his short chapter does not mention Durham. Treasure-trove was yet another matter dealt with by procurators fiscal in Scotland. A. R. G. McMillan, *The law of bona vacantia in Scotland* (Edinburgh, 1936).

13   Durham University Archives DHC4/194607, Chester coroner's rentals, 1622–1761. DHC4/191623–191649, Easington coroner's rentals, 1619–1784. R. L. Storey, *Thomas Langley and the bishopric of Durham, 1406–1437* (London: SPCK, 1961), 69–70. C. D. Liddy, *The bishopric of Durham in the late Middle Ages: lordship, community and the cult of St Cuthbert* (Woodbridge: Boydell, 2008), 155–61.

14   Storey, *Thomas Langley*, 62. G. T. Lapsley, *The county palatine of Durham: a study in constitutional history* (London: Longmans, Green, 1900), 87. Neville, 'Office of coroner', 54. Lloyd, *A history of Carmarthenshire*, vol. 1, 218–19. Rees, 'Survivals of ancient Celtic custom', 164.

15   M. Holford, 'Durham under Bishop Anthony Bek, 1283–1311', in M. L. Holford and K. J. Stringer (eds), *Border liberties and loyalties: north-east*

England, c. 1200–c. 1400 (Edinburgh: Edinburgh UP, 2010), 138–9, 148–9. R. H. Britnell, 'Feudal reaction after the Black Death in the palatinate of Durham', *Past & Present* 128 (1990), 32–3, 41.

16   W. R. B. Robinson, 'The government of the lordship of Gower and Kilvey in the early-Tudor period', in T. B. Pugh (ed.), *Glamorgan county history, volume III: the Middle Ages* (Cardiff: University of Wales Press, 1971), 267.

17   Hunnisett, 'Origins', 99. Stewart-Brown, *Serjeants of the peace*, 63–5. E. G. Kimball, *Serjeanty tenure in medieval England* (New Haven: Yale UP, 1936), 84–6. Richardson and Sayles, *Governance of mediaeval England*, 424. For a Lancashire example of a holding 'per serjenciam custod. eyras regias' see T. D. Whitaker, *An history of Richmondshire, in the North Riding of the country of York* 2 vols (London: Longman, Hurst, Rees, Orme, and Browne, 1823), vol. 2, 474. Serjeanty was a 'feudal tenure on condition of rendering some specified personal service to the king' (OED). English serjeants were mostly appointed by sheriffs or franchise-holders and early serjeanty could be a base tenure. S. H. Rigby, *English society in the later Middle Ages* (Basingstoke: Macmillan, 1995), 39. R. Faith, *The English peasantry and the growth of lordship* (London: Leicester UP, 1997), 95, 240.

18   Innes, *Scotch legal antiquities*, 62. ERS XIV, 628. Robertson, *Missing charters*, 'Serjeandry, office of'. *Protocol books of James Foulis, 1546–1553, and Nicol Thounis, 1559–1564* Scottish Record Society 57 (Edinburgh, 1927), Nicol Thounis, no. 202. Hunnisett, 'Origins', 100. However, see also Walker, *Legal history of Scotland* vol. 1, 81, 158, 159.

19   Robertson, *Sheriff Courts*, 5.

20   Duke of Rutland, *Journal of a tour through north and south Wales, the Isle of Man, etc.* (London: J. Triphook, 1805), 256, 261. D. Craine, *Manannan's Isle* (Glasgow: Manx Museum and National Trust, 1955), 57–62. J. R. Dickinson, *The lordship of Man under the Stanleys: government and economy in the Isle of Man, 1580–1704* Chetham Society 3rd series 41 (Manchester, 1996), 48.

21   *Atlas of Scottish history to 1707*, 190. G. W. S. Barrow, *Scotland and its neighbours in the Middle Ages* (London: Hambledon, 1992), 221.

22   A. W. Moore, *A history of the Isle of Man* (London: T. Fisher Unwin, 1900), 746, 800. T. Talbot (ed.), *The manorial roll of the Isle of Man, 1511–1515* (London: Oxford UP, 1924), 58, 61, 66.

23   Skene, *Celtic Scotland*, vol. 3, 279–80. Skene (ed.), *Fordun's chronicle*, vol. 2, 458–9. A. Ashley, 'The spiritual courts of the Isle of Man, especially in the seventeenth and eighteenth centuries', *English Historical Review* 72 (1957), 54. J. Sharpe, 'Towards a legal anthropology of the early modern Isle of Man', in R. McMahon (ed.), *Crime, law and popular culture in Europe, 1500–1900* (Cullompton: Willan Publishing, 2008), 123.

24   Rutland, *Journal of a tour*, 261. Talbot (ed.), *Manorial roll of the Isle of Man*, 5, 9, 22, 25, 34, 39, 47, 50, 54, 58, 61, 65, 70, 75, 76.

DOI: 10.1057/9781137381071.0010

25 Moore, *Isle of Man*, 746.
26 N. Denholm-Young, *Seignorial administration in England* (London: Oxford UP, 1937), 105. Braddick, *State formation*, 27–46.
27 G. Poulson, *The history and antiquities of the seigniory of Holderness* 2 vols (Hull: R. Brown, 1840–1), vol. 1, 158.
28 The last Percy Earl died in 1670. His daughter and heiress, Elizabeth, after two brief childless marriages, married Charles Seymour, Duke of Somerset in 1682. Egremont was part of the extensive Percy estate in West Cumberland supervised by a single auditor and known as the Honour of Cockermouth.
29 Lapsley, *County palatine*, 86–8. Hunnisett, *Medieval coroner*, 95. D. Loades, *Tudor government: structures of authority in the sixteenth century* (Oxford: Blackwell, 1997), 217–19. P. L. Larson, *Conflict and compromise in the late medieval countryside: lords and peasants in Durham, 1349–1400* (London: Routledge, 2006), 56–7. At some periods, the coronership of Chester ward was also hereditary: it was in the hands of the Lumley family for generations until the Pilgrimage of Grace. ODNB 'John Lumley, fifth baron Lumley'. Some thirteenth-century Northumberland coroners were hereditary serjeanty tenants. Kimball, *Serjeanty*, 84–6. Cheshire was the only other area with a hereditary serjeant and there were apparently no hereditary coroners outside Durham. Stewart-Brown, *Serjeants of the peace*, 73. However, Jewell, *English local administration*, 156, notes that they existed in Northumberland 1194–1246.
30 W. H. D. Longstaffe, *The history and antiquities of the parish of Darlington* (London: J. H. Parker, 1854), 278.
31 M. Bonney, *Lordship and the urban community: Durham and its overlords, 1250–1540* (Cambridge: Cambridge UP, 1990), 209, 211.
32 Huntington Library, Hastings Ham box 2/1, pp. 10, 18–19. R. Evans, 'Merton College's control of its tenants at Thorncroft, 1270–1349', in Z. Razi and R. Smith (eds), *Medieval society and the manor court* (Oxford: Clarendon, 1996), 210–20.
33 Cumbria Archives Service Ca/2/390.
34 R. W. Hoyle (ed.), 'Letters of the Cliffords, lords Clifford and earls of Cumberland, c.1500–c.1565', *Camden Miscellany XXXI* (London: Royal Historical Society, 1992), 93.
35 L. B. Smith, 'A contribution to the history of *galanas* in late-medieval Wales', *Studia Celtica* 43 (2009), 87–94. Ellis, 'British perspective', 62. Woolf, *Pictland to Alba*, 346–9.
36 G. W. Bernard, *The power of the early Tudor nobility: a study of the fourth and fifth earls of Shrewsbury* (Brighton: Harvester, 1985), 180. T. Thornton, 'Local equity jurisdictions in the territories of the English crown: the palatinate of Chester, 1450–1540', in D. E. S. Dunn (ed.), *Courts, counties and the capital in the later Middle Ages* (Stroud: Sutton, 1996), 27–52, argues against this assumption.

DOI: 10.1057/9781137381071.0010

37   Cumbria Archives Service D/LEC/CRI/1/2. Impey, *Office of sheriff*, 11–15, 111–14.

38   Cumbria Archives Service D/Lons/L1/1/24/24: letter from William Smith, Bowerbank, Penrith, 10 October 1673.

39   University of Nottingham Library, Pw F 3211 (17 March 1763). G. H. H. Glasgow, 'The election of county coroners in England and Wales *circa* 1800–1888', *Journal of Legal History* 20, 3 (1999), 75–108. Eastwood, *Governing rural England*, 67–8.

40   Kimball, *Serjeanty*, 84–6, 203.

41   R. Muir and E. M. Platt, *A history of municipal government in Liverpool from the earliest times to the Municipal Reform Act of 1835* (London: Williams & Norgate, 1906), 144. This may later have been regularised. *Ibid.*, 281, 283–6. J. A. Twemlow (ed.), *Liverpool town books: proceedings of assemblies, common councils, portmoot books etc, 1550–1862* [*sic.*: 1550–1603] 2 vols (Liverpool: Constable, 1918–35), vol. 2, 766n. See also Nelson, *Justice of Peace*, 197–200. Gross (ed.), *Select cases*, xx and n, suggests that thirteenth- and fourteenth-century deputies were scribes or *clerici*.

42   S. K. Roberts (ed.), *Evesham borough records of the seventeenth century, 1605–1687* (Kendal: Titus Wilson, 1994), index 'coroners, elected'.

43   [Fitzherbert] *Offices of Shyriffes* [np]. Hunnisett, *Medieval coroner*, 190–1, notes that Durham coroners could deputise and this may have been general with franchisal appointments, where the holder was strictly speaking the Lord. J. Eaton, 'Sketch of the coroner's court, and its principal relations to the medical profession', *The Provincial Medical Journal* 7 (1888), 252. By 6 and 7 Vic., c. 83 (1843) county coroners were authorised to appoint deputies subject to the approval of the Lord Chancellor.

44   RMS V, no. 2001. APS V, 98b. P. J. Anderson (ed.), *Charters and other writs ... of the royal burgh of Aberdeen, 1171–1804* (Aberdeen: Aberdeen Town Council, 1890), 183–4. Maidment (ed.), *Chronicle of Perth*, 31–2. Walker, *Legal history of Scotland* vol. 4, 467. Gross (ed.), *Select cases*, xxiii. S. and B. Webb, *English local government from the revolution to the municipal corporations act: the manor and the borough* (London: Longman, Green and co., 1924), index 'coroners, borough'.

45   Perth & Kinross Archives B59/26/1/6 and 134, confirmed by a charter of 15 November 1600.

46   A. D. M. Barrell, R. R. Davies, O. J. Padel and L. B. Smith, 'The Dyffryn Clwyd court roll project, 1340–1352 and 1389–1399: a methodology and some preliminary findings', in Z. Razi and R. Smith (eds), *Medieval society and the manor court* (Oxford: Clarendon, 1996), 279–80. The coroners here, new in the mid-fourteenth century, were mainly concerned with the death of the Lord's livestock and so were also called *cadavatores*.

47   Stewart-Brown, *Serjeants of the peace*, 2. N. Sellers, 'Tenurial serjeants', *American Journal of Legal History* 14 (1970), 319–32. [Fitzherbert] *Offices of Shyriffes* [np].

DOI: 10.1057/9781137381071.0010

48 Stewart-Brown, *Serjeants of the peace*, 64–5. R. B. Pugh, *Imprisonment in medieval England* (Cambridge: Cambridge UP, 1968), 280–6.
49 NA STAC 2/18/222. See also STAC 2/17/188. R. Stewart-Brown, *The wapentake of Wirral: a history of the royal franchise of the hundred and hundred court of Wirral in Cheshire* (Liverpool: Young, 1907). Yeomen of the king were also active in making arrests around this time. NA STAC 2/19/158.
50 Thornton, *Cheshire*, 131–2.
51 R. N. Swanson, *Church and society in late medieval England* (Oxford: Blackwell, 1989), 136-9.
52 Braddick, *State formation*, 352.
53 Hunnisett, *Medieval coroner*, 198, notes a statute of 1368 requiring an escheator to have land in fee worth £20 a year. With the English Statute of Tenures of 1660, escheators and feodaries effectively became defunct.
54 Hunnisett, 'Origins', 98–9.
55 Clayton, *Administration of Chester*, 190.
56 Liddy, *Bishopric of Durham*, 156–61. Neville, 'Office of coroner', 54.
57 Holford, 'Durham under Bishop Anthony Bek', 138–9, 148–9. It should, however, be noted that the master forester and coroner of Chester ward had been in the family of the lords Lumley during the fifteenth century and John, Lord Lumley was only replaced by Sir William Eure in 1524. James, *Family, lineage, and civil society*, 44 and n.
58 Hunnisett, *Medieval coroner*, 192.
59 K. Stringer, 'Tynedale: power, society and identities, c.1200–1296', in M. L. Holford and K. J. Stringer (eds), *Border liberties and loyalties: north-east England, c. 1200–c. 1400* (Edinburgh: Edinburgh UP, 2010), 236, 272–3, 282. K. Stringer, 'Tynedale: a community in transition, 1296–c.1400', in *ibid.*, 316. Hexhamshire coroners too were elective in theory. M. Holford, 'Hexhamshire and Tynemouthshire', in *ibid.*, 192, 194, 210. The special administrative status of Tyndedale and Redesdale was not ended until 1495/6. S. J. Watts, *From border to middle shire: Northumberland, 1586–1625* (Leicester: Leicester UP, 1975), 24. The same short-term appointments are suggested in part of mid-fourteenth-century north Wales. Barrell *et al.*, 'Court roll project', 279–80. Redesdale and Tynedale remained for much of the Tudor period semi-independent franchises under military wardens. Loades, *Tudor government*, 81–2. C. M. Newman, 'Local court administration within the liberty of Allertonshire, 1470–1540', *Archives* 22 (1995), 13–24. Ellis, 'British perspective', 56–60. K. J. Kesselring, ' "Berwick is our England": Local and national identities in an Elizabethan border town', in D. Woolf and N. L. Jones (eds), *Local identities in late medieval and early modern England* (London: Palgrave Macmillan, 2007), 92–112.

Conclusion: Coroners and British history

Abstract: *The conclusion highlights the importance of regional experience, seen through the lens of the work of coroners, to understanding Britain's national historical development, especially changes in the personnel and workings of the law. It argues that a balance between Scottish and English perspectives has much to offer legal, social and political historians. The book provides essential historical background for understanding the work of coroners in the modern world.*

Houston, R. A. *The Coroners of Northern Britain c. 1300–1700*. Basingstoke: Palgrave Macmillan, 2014. DOI: 10.1057/9781137381071.0011.

In 1702 Blackerby Fairfax, an English physician, published a pro-Union tract, which claimed, among much else that ranged from the tendentious to the mendacious, close similarities in law and its officers between Scotland and England. 'We have', he trumpeted, 'the same Ministers of Justice, as Sherffs [sic], Coroners, &c.'¹ Although he correctly identified the presence of these officers, Fairfax knew little of Scotland or he would have seen how utterly different both named offices were at the time he wrote. His Unionist perspective, which has also deeply influenced work on other Scottish officers like Justices of Peace, is one of the foundations of a modern historiography that believes the development of institutions should be judged by how closely they approximate to (south-east) English equivalents.² Instead, the history of Scotland's coroners and the investigation of sudden or suspicious deaths illuminate the diverse political relationships between centre and locality in the component parts of Britain. They show the enduring significance of regional as well as national legal cultures to the centralisation, regularisation or extensification of government, which took place in very different ways in the diverse parts of the British Isles. Above all, they illustrate the importance of Maitland's modest recognition of 'that fatal disease of contented insularity which so easily besets' the English and his conclusion that 'there is nothing that sets a man thinking and writing to such good effect about a system of law and its history as an acquaintance however slight with other systems and their history'.³

For one thing Continental and Scottish magistrates both investigated sudden deaths privately and at discretion, to resolve whether a crime had likely been committed, where English coroners had immediately to hold public inquests to allay local suspicions (perhaps also to appease or 'provide closure' for the family of the dead person) and to determine exact cause of death, even in cases beyond suspicion. Some late-eighteenth-century proponents of Scottish criminal law reform, like the advocate Hugo Arnot, decried the lack of routine inquests and there were occasional calls for the creation of a Scottish equivalent.⁴ As advocate-depute Archibald Alison put it with feeling in 1825: 'It is a remarkable fact, that, while the English are the people in the world who are most firmly attached to the institutions of their own ancestors ... They resist in the most strenuous manner any attempt to introduce an alteration in their own judicial proceedings, but they lend a ready ear to any person who proposes similar alterations upon the laws of any other people'.⁵ In contrast, certain Victorian observers made a positive virtue of the

DOI: 10.1057/9781137381071.0011

discretion and secrecy of Scottish investigative procedure, asserting that 'the privacy, which forms such an essential feature in all Scots criminal procedure, is maintained. The police reports are private, the witnesses are examined privately, and outwith the presence of each other, and the reports by the procurator-fiscal and opinions of counsel thereon are also confidential.'[6] Writing in 1855, the Scottish surgeon James Craig thought English crowner law 'lauded by a few and condemned by many. It is a relic of a bygone age.'[7]

Privacy is not solely a sign of care. Historian Richard Smith suggests that the discretion of Roman practice distinguished it from the traditions of centralisation and public involvement that had characterised England since the Middle Ages. 'Secret interrogation may well have been a more likely development in societies in which a "participatory" tradition in policing and trial of "crime" was poorly developed or only weakly related to the instructions sent out from central government.'[8] Historian Julian Goodare has shown the weak links between centre and locality in early modern Scottish government, for example in failed attempts to create civil parishes on the English model; nor were manors introduced during the Middle Ages, both probably because baronies performed many of the functions.[9] The way the coronership developed in Scotland demonstrates the growing power of the crown, but also the structural limits on change until the early modern period, as for centuries Scottish monarchs accommodated the diverse institutional frameworks and social priorities of its component peoples. By c.1200 criminal justice belonged largely to English kings, whereas in Scotland the king delegated justice much more extensively to particular local nobles until centuries later.

If central control was dispersed and shared in Scotland, local participation too was understood differently. Scotland's people had a part to play in enforcing laws, but they were not involved so directly as to vote for coroners – a provision introduced by the English crown to reduce grievances and to give electors (and jurors) a direct stake in properly exercising the office.[10] Blackstone noted the 'ancient' (i.e. late medieval) practice of electing all local officials who might intrude on subjects' liberties.[11] Under the tithing system, hundreds and vills (*trefydd* in Wales) were responsible for unreported felonies and for not presenting indicted criminals, whereas it was the sheriff and/or justiciar in Scotland.[12] Late medieval and early modern Scottish government could not match the capacity of its English counterpart 'to devise and comprehensively institute neighbourhood-level institutions' for a wide range of administrative

DOI: 10.1057/9781137381071.0011

tasks.[13] Where the English coroner was one of many interfaces between government and people, his Scottish equivalent was a direct instrument of the crown. Scotland had juries of trial (assizes) and, in civil matters, inquiry (inquests), but not really of presentment and indictment in criminal matters, limiting the participation of the relatively humble in local government, except at the very lowest level of jurisdiction.[14] We have seen how constables and JPs, the cornerstone of peace-keeping in England's localities, were much less important in Scotland until the eighteenth century. Instead, individual lords operated mechanisms like the 'general band', a bond that landowners would be responsible for their tenants and followers.[15] Until the Tudor era, the power of franchisal coroners appointed by local lords also distinguished much of the north of England from the combination of strong central authority and respon-siveness to pressures from below, characteristic of the rest of England.[16] Scotland's polity was different from England's and the processes by which government became regularised, although broadly similar, were sufficiently distinctive (not least in chronology) to make it impossible to speak of a single British model of political and administrative change.

The coroner in England was and is an 'independent judicial officer, who is solely responsible, subject to the requirements of the law, for the conduct of his duties'; the Royal Commission on the County Rates (1834–6) described him as a 'judicial functionary'.[17] The coroner in Scotland was an executive judicial officer who enforced the king's (com-mon) law on felons, appointed by the crown or its nominee and always subordinate to and increasingly subsumed within the growing power of the sheriff. With largely ancillary and extra-curial functions he looks more like a twelfth-century English sheriff's serjeant or a keeper of the peace than a later English coroner.[18] The English coroner was a check on the sheriff whereas his Scottish counterpart could be simultaneously (in title at least) sheriff and coroner of the same jurisdiction; the actual officers were expected to work together, though the coroner could oper-ate alone.[19] Justiciars supervised Scottish sheriffs while ayres were active, though Justices of Peace were also meant to act as a curb on the increas-ingly powerful sheriffs after 1587.

Scottish coroners investigated some suspicious deaths, but their job focused more on the living and especially on citing offenders and seizing their goods.[20] They never seem to have performed any properly judicial role either alone or with sheriffs or justiciars, except perhaps a measure of summary justice, and they were subject to the supervision of both magnates and men of law. Those who seem to have had greater powers acted by virtue of some other title such as forester or steward, the latter another once-high office that declined in status during the fifteenth century.[21] Tudor and later English coroners specialised in investigating sudden death whereas Scottish procurators fiscal were generalists. As part of their routine business, Scottish magistrates carried out tasks allocated to English coroners. For example, ordinary officers (usually the ones passing the sentence) administered banishment from Scotland (the equivalent of English abjuration of the realm) or from a jurisdiction within it. Finally, the task of English coroners became increasingly burdensome after 1750, where the Scottish coroner's workload had faded to nothing by that date.[22]

Notes

1   B. Fairfax, *A discourse upon the uniting Scotland with England: containing the general advantage of such an union to both Kingdoms* (London: J. Knapton, 1702), 54.

2   Moir, 'Justices of the peace', 12–14.

3   F. W. Maitland, 'The laws of the Anglo-Saxons', in H. A. L. Fisher (ed.), *The collected papers of Frederic William Maitland*, 3 vols (Cambridge, 1911), vol. 3, 460; 'Why the history of English law is not written', in *ibid.*, vol. 1, 489.

4   H. Arnot, *A collection and abridgement of celebrated criminal trials in Scotland, from A.D. 1536, to 1784* (Edinburgh: W. Smellie, 1785), 236. 'Law cases and speculative questions for the discussion of the Scots Law Society during session [March 1827]', 'Would it be expedient to introduce the practice of the coroner's inquests into this country?' [Edinburgh UL C.R.9.4.32/24, p. 4]. There is no transcript of the debate.

5   Alison, *Criminal justice*, 1.

6   Renton, 'Sudden death in Scotland', 168. Russell, 'Procedure in criminal prosecutions', 261. Anon., 'Official inquiry in cases of sudden death', *Journal of Jurisprudence* 2 (1858) 271–6.

7   Craig, *Coroner*, 16.

8   R. M. Smith, '"Modernization" and the corporate medieval village community in England: some sceptical reflections', in A. R. H. Baker and D. Gregory (eds), *Explorations in historical geography: interpretative essays* (Cambridge: Cambridge UP, 1984), 171.

9   Goodare, *Government of Scotland*, 192–219, esp. p. 202. Grant, 'Franchises north of the border', 161, 164–6. Ellis, 'British perspective', 56.

10 Gross (ed.), *Select cases*, xxvii.

DOI: 10.1057/9781137381071.0011

11 Blackstone, *Commentaries*, I.9.II [vol. 1, 335].

12 B. H. Putnam (ed.), *Kent keepers of the peace, 1316–1317* (London, 1933), xxix. Stewart-Brown, *Serjeants of the peace*, 33–46. N. M. W. Powell, 'Crime and criminality in Denbighshire during the 1590s: the evidence of the records of the Great Sessions', in J. G. Jones (ed.), *Class, community and culture in Tudor Wales* (Cardiff: University of Wales Press, 1989), 261–94. Hunnisett, *Medieval coroner*, 27–8, notes kin standing surety. In much of England, sheriffs supervised the frankpledge system through their twice yearly tourns. The absence of tourns from the north may be because the peace-keeping system was different.

13 J. Innes, 'The state of the poor in eighteenth-century England in European perspective', in J. Brewer and E. Hellmuth (eds), *Rethinking Leviathan: the eighteenth-century state in Britain and Germany* (Oxford: Oxford UP, 1999), 243.

14 Willock, *Jury*, 25, 144–7.

15 Goodare, *State and society*, 258–9. In Tynedale (Northumberland) the system was called 'booking'. Ellis, 'British perspective', 57. Henry VII tried something similar in Wales with 'indentures of the Marches' to deal with perversion of a once-laudable custom called *arddel*, where lords employed outlaws in their private armies; Henry VIII passed legislation banning the practice. J. G. Jones, *Early modern Wales, c.1525–1640* (Basingstoke, 1994), 55, 72–3. In earlier Cheshire, the system of avowry seems to have been based on Welsh practices. Stewart-Brown, *Serjeants of the peace*, 3.

16 Ellis, 'British perspective', 56.

17 'Brodrick Report', 120. PP, 1836, XXVII (58), 38.

18 Hunnisett, 'Origins', 96–7.

19 *Historical notices*, 624. Gross (ed.), *Select cases*, xxv–vii. Cam, 'Shire officials', 156.

20 Skene, *De Verborum Significatione*, 'porteous', 'traistis'. The traistis were catalogues of charges.

21 Gilbert, *Hunting*, 147. Murray, 'Administrators of church lands', 40–1.

22 Eastwood, *Governing rural England*, 68–9.

DOI: 10.1057/9781137381071.0011

Select Bibliography of Works Published since 1900

An introduction to Scottish legal history Stair Society vol. 20 (Edinburgh: R. Cunningham, 1958).

Brown, J. M. (ed.), *Scottish society in the fifteenth century* (London: Edward Arnold, 1977).

Brown, M., 'Scotland tamed? Kings and magnates in late medieval Scotland: a review of recent work', *Innes Review* 45 (1994), 120–46.

Cairns, J. W., 'Historical introduction', in K. Reid and R. Zimmermann (eds), *A history of private law in Scotland. Volume 1: introduction and property* (Oxford: Oxford UP, 2000), 14–184.

Carmichael, I. H. B., *Sudden deaths and fatal accident inquiries: Scots law and practice* (1986. 2nd edition, W. Green: Edinburgh, 1993).

Clark, M. and Crawford, C. (eds), *Legal medicine in history* (Cambridge: Cambridge UP, 1994).

Davies, S. J., 'The courts and the Scottish legal system 1600–1747: the case of Stirlingshire', in V. A. C. Gatrell, B. Lenman and G. Parker (eds), *Crime and the law: the social history of crime in western Europe since 1500* (London: Europa, 1980), 120–54.

Death and the procurator fiscal (Edinburgh: Crown Office, 1998).

Dickinson, W. C., *Scotland from the earliest times to 1603* (1961. 3rd edition revised and edited by A. A. M. Duncan. Oxford: Oxford UP, 1977).

DOI: 10.1057/9781137381071.0012

Emsley, K. and Fraser, C. M., *The courts of the palatinate of Durham from earliest times to 1971* (Newcastle: Pattinson and sons, 1984).

Farrell, B., *Coroners: practice and procedure* (Dublin: Round Hall, 2000).

Finlay, J., *Men of law in pre-reformation Scotland* (East Linton: Tuckwell, 2000).

Forte, A. D. M., 'The horse that kills: some thoughts on deodands, escheats and crime in fifteenth century Scots law', *Tijdschrift voor Rechtsgeschiedenis* 58 (1990), 95–110.

Gillies, W., 'Some thought on the Toschederache', *Scottish Gaelic Studies* 17 (1996), 128–42.

Godfrey, A. M., *Civil justice in Renaissance Scotland: the origins of a central court* (Leiden: Brill, 2009).

Goodare, J., *The government of Scotland, 1560–1625* (Oxford: Oxford UP, 2004).

Grant, A. and Stringer, K. J. (eds), *Medieval Scotland: crown, lordship and community* (Edinburgh: Edinburgh UP, 1993).

Gunn, S. J., *Early Tudor government, 1485–1558* (Basingstoke: Macmillan, 1995).

Harding, A., 'The origins and early history of the keepers of the peace', *Transactions of the Royal Historical Society* 5th series 10 (1960), 85–109.

——. *A social history of English law* (Harmondsworth: Penguin, 1966).

Holford, M. L. and Stringer, K. J. (eds), *Border liberties and loyalties: north-east England, c. 1200–c. 1400* (Edinburgh: Edinburgh UP, 2010).

Houston, R. A., *Punishing the dead? Suicide, lordship, and community in Britain, 1500–1830* (Oxford: Oxford UP, 2010).

Hunnisett, R. F., 'The origins of the office of coroner', *Transactions of the Royal Historical Society* 5th series 8 (1958), 85–104.

——. *The medieval coroner* (Cambridge: Cambridge UP, 1961).

Kingston, S., *Ulster and the Isles in the fifteenth century: the lordship of the Clann Domhnaill of Antrim* (Dublin: Four Courts, 2004).

Liddy, C. D., *The bishopric of Durham in the late Middle Ages: lordship, community and the cult of St Cuthbert* (Woodbridge: Boydell, 2008).

MacQueen, H. L., 'The laws of Galloway: a preliminary survey', in R. D. Oram and G. P. Stell (eds), *Galloway: land and lordship* (Edinburgh: Scottish Society for Northern Studies, 1991), 131–43.

——. *Common law and feudal society in medieval Scotland* (Edinburgh: Edinburgh UP, 1993).

DOI: 10.1057/9781137381071.0012

Márkus, G., 'Dewars and relics in Scotland: some clarifications and questions', *Innes Review* 60 (2009), 95–144.

Murray, P. J., 'The lay administrators of church lands in the fifteenth and sixteenth centuries', *Scottish Historical Review* 74, 197 (1995), 26–44.

Neville, C. J., ' "The bishop's ministers": the office of coroner in late medieval Durham', *Florilegium* 18, 2 (2001), 47–60.

Prestwich, M. (ed.), *Liberties and identities in the medieval British Isles* (Woodbridge: Boydell, 2008).

Rees, W., 'Survivals of ancient Celtic custom in medieval England', in H. Lewis (ed.), *Angles and Britons: O'Donnell lectures* (Cardiff: University of Wales Press, 1963), 148–68.

Sellar, W. D. H., 'Celtic law and Scots law: survival and integration', *Scottish Studies* 29 (1989), 1–27.

Sellers, N., 'Tenurial serjeants', *American Journal of Legal History* 14 (1970), 319–32.

Sharpe, J. and Dickinson, J. R., 'Coroners' inquests in an English county, 1600–1800: a preliminary survey', *Northern History* 48 (2011), 253–69.

Sheehan, A. V., *Criminal procedure in Scotland and France* (Edinburgh: HMSO, 1975).

Stevenson, K. (ed.), *The herald in late medieval Europe* (Woodbridge: Boydell, 2009).

Stewart-Brown, R., *The serjeants of the peace in medieval England and Wales* (Manchester: Manchester UP, 1936).

Thornton, T., *Cheshire and the Tudor state, 1480–1560* (Woodbridge: Boydell, 2000).

Tompson, R. S., 'The justices of the peace and the United Kingdom in the age of reform', *Journal of Legal History* 7 (1986), 273–92.

Walker, D. M., *A legal history of Scotland* 7 vols (Edinburgh: W. Green etc., 1988–2004).

Wellington, R. H., *The king's coroner* (London: W. Clowes, 1905).

DOI: 10.1057/9781137381071.0012

Index

DOI: 10.1057/9781137381071.0013

DOI: 10.1057/9781137381071.0013